W9-DDK-087

SKILLS MASTERY

This Book Includes:

- Practice questions to help students master topics assessed on the the PARCC and Smarter Balanced Tests.
  - ▸ The Number System
  - ▸ Expressions & Equations
  - ▸ Functions
  - ▸ Geometry
  - ▸ Statistics & Probability
- Detailed Answer explanations for every question
- Strategies for building speed and accuracy
- Content aligned with the Common Core State Standards

Plus access to Online Workbooks which include:

- Hundreds of practice questions
- Self-paced learning and personalized score reports
- Instant feedback after completion of the workbook

Complement Classroom Learning All Year

Using the Lumos Study Program, parents and teachers can reinforce the classroom learning experience for children. It creates a collaborative learning platform for students, teachers and parents.

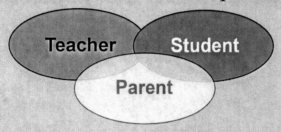

Used in Schools and Libraries
To Improve Student Achievement

Lumos Learning

Common Core Practice - Grade 8 Math: Workbooks to Prepare for the PARCC or Smarter Balanced Test

Contributing Author   -   Nicole Fernandez
Contributing Author   -   Nancy Chang
Curriculum Director   -   Marisa Adams
Executive Producer   -   Mukunda Krishnaswamy
Designer and Illustrator -   Sowmya R.

ISBN-10: 1940484480

ISBN-13: 978-1-940484-48-8

Printed in the United States of America

For permissions and additional information contact us

Lumos Information Services, LLC
PO Box 1575, Piscataway, NJ 08855-1575
http://www.LumosLearning.com

Email: support@lumoslearning.com
Tel: (732) 384-0146
Fax: (866) 283-6471

Lumos Learning

Table of Contents

Introduction

The Common Core State Standards Initiative (CCSS) was created from the need to have more robust and rigorous guidelines, which could be standardized from state to state. These guidelines create a learning environment where students will be able to graduate high school with all skills necessary to be active and successful members of society, whether they take a role in the workforce or in some sort of post-secondary education.

Once the CCSS were fully developed and implemented, it became necessary to devise a way to ensure they were assessed appropriately. To this end, states adopting the CCSS have joined one of two consortia, either PARCC or Smarter Balanced.

Why Practice by Standard?

Each standard, and substandard, in the CCSS has its own specific content. Taking the time to study and practice each one individually can help students more adequately understand the CCSS for their particular grade level. Additionally, students have individual strengths and weaknesses. Being able to practice content by standard allows them the ability to more deeply understand each standard and be able to work to strengthen academic weaknesses.

How Can the Lumos Study Program Prepare Students for Standardized Tests?

Since the fall of 2014, student mastery of Common Core State Standards are being assessed using standardized testing methods. At Lumos Learning, we believe that yearlong learning and adequate practice before the actual test are the keys to success on these standardized tests. We have designed the Lumos study program to help students get plenty of realistic practice before the test and to promote yearlong collaborative learning.

This is a Lumos tedBook™. It connects you to Online Workbooks and additional resources using a number of devices including android phones, iPhones, tablets and personal computers. Each Online Workbook will have some of the same questions seen in this printed book, along with additional questions. The Lumos StepUp® Online Workbooks are designed to promote yearlong learning. It is a simple program students can securely access using a computer or device with internet access. It consists of hundreds of grade appropriate questions, aligned to the new Common Core State Standards. Students will get instant feedback and can review their answers anytime. Each student's answers and progress can be reviewed by parents and educators to reinforce the learning experience.

How to use this book effectively

The Lumos Program is a flexible learning tool. It can be adapted to suit a student's skill level and the time available to practice before standardized tests. Here are some tips to help you use this book and the online resources effectively:

Students

- The standards in each book can be practiced in the order designed, or in the order of your own choosing.
- Complete all problems in each workbook.
- Use the Online workbooks to further practice your areas of difficulty and complement classroom learning.
- Download the Lumos StepUp® app using the instructions provided to have anywhere access to online resources.
- Practice full length tests as you get closer to the test date.
- Complete the test in a quiet place, following the test guidelines. Practice tests provide you an opportunity to improve your test taking skills and to review topics included in the CCSS related standardized test.

Parents

- Familiarize yourself with your state's consortium and testing expectations.
- Get useful information about your school by downloading the Lumos SchoolUp™ app. Please follow directions provided in "How to download Lumos SchoolUp™ App" section of this chapter.
- Help your child use Lumos StepUp® Online Workbooks by following the instructions in "How to access the Lumos Online Workbooks" section of this chapter.
- Help your student download the Lumos StepUp® app using the instructions provided in "How to download the Lumos StepUp® Mobile App" section of this chapter.
- Review your child's performance in the "Lumos Online Workbooks" periodically. You can do this by simply asking your child to log into the system online and selecting the subject area you wish to review.
- Review your child's work in each workbook.

Teachers

- You can use the Lumos online programs along with this book to complement and extend your classroom instruction.

- Get a Free Teacher account by using the respective states specific links and QR codes below:

PARCC States	SBAC States
LumosLearning.com/a/stepupbasic	LumosLearning.com/a/sbacbasic

This Lumos StepUp® Basic account will help you:

- Create up to 30 student accounts.
- Review the online work of your students.
- Easily access CCSS.
- Create and share information about your classroom or school events.

NOTE: There is a limit of one grade and subject per teacher for the free account.

- Download the Lumos SchoolUp™ mobile app using the instructions provided in "How can I Download the App?" section of this chapter.

How to Access the Lumos Online Workbooks

First Time Access:

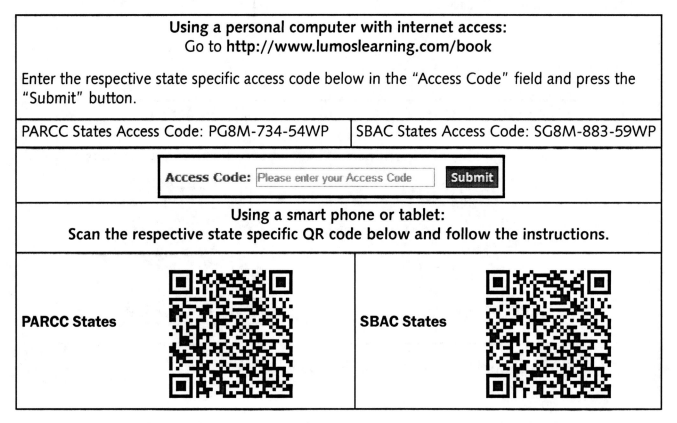

Using a personal computer with internet access:
Go to http://www.lumoslearning.com/book

Enter the respective state specific access code below in the "Access Code" field and press the "Submit" button.

| PARCC States Access Code: PG8M-734-54WP | SBAC States Access Code: SG8M-883-59WP |

Access Code: [Please enter your Access Code] **Submit**

Using a smart phone or tablet:
Scan the respective state specific QR code below and follow the instructions.

PARCC States

SBAC States

In the next screen, click on the "New User" button to register your user name and password.

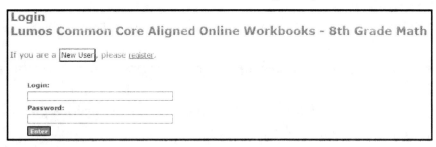

Login
Lumos Common Core Aligned Online Workbooks - 8th Grade Math

If you are a [New User], please register.

Login:
_____

Password:
_____

[Enter]

Subsequent Access:

After you establish your user id and password for subsequent access, simply login with your account information.

What if I buy more than one Lumos Study Program?

Please note that you can use all Online Workbooks with one User ID and Password. If you buy more than one book, you will access them with the same account.

Go back to the **http://www.lumoslearning.com/book** link and enter the access code provided in the second book. In the next screen simply login using your previously created account.

Lumos StepUp® Mobile App FAQ For Students

What is the Lumos StepUp® App?

It is a FREE application you can download onto your Android smart phones, tablets, iPhones, and iPads.

What are the Benefits of the StepUp® App?

This mobile application gives convenient access to Practice Tests, Common Core State Standards, Online Workbooks, and learning resources through your smart phone and tablet computers.

- Eleven Technology enhanced question types in both MATH and ELA
- Sample questions for Arithmetic drills
- Standard specific sample questions
- Instant access to the Common Core State Standards
- Jokes and cartoons to make learning fun!

Do I Need the StepUp® App to Access Online Workbooks?

No, you can access Lumos StepUp® Online Workbooks through a personal computer. The StepUp® app simply enhances your learning experience and allows you to conveniently access StepUp® Online Workbooks and additional resources through your smart phone or tablet.

How can I Download the App?

Visit **lumoslearning.com/a/stepup-app** using your smart phone or tablet and follow the instructions to download the app.

QR Code
for Smart Phone
Or Tablet Users

Lumos SchoolUp™ Mobile App FAQ For Parents and Teachers

What is the Lumos SchoolUp™ App?

It is a FREE App that helps parents and teachers get a wide range of useful information about their school. It can be downloaded onto smartphones and tablets from popular App Stores.

What are the Benefits of the Lumos SchoolUp™ App?

It provides convenient access to

- School "Stickies". A Sticky could be information about an upcoming test, homework, extra curricular activities and other school events. Parents and educators can easily create their own sticky and share with the school community.
- Common Core State Standards.
- Educational blogs.
- StepUp™ student activity reports.

How can I Download the App?

Visit **lumoslearning.com/a/schoolup-app** using your smartphone or tablet and follow the instructions provided to download the App. Alternatively, scan the QR Code provided below using your smartphone or tablet computer.

QR Code
for Smart Phone
Or Tablet Users

The Number System

Rational vs. Irrational Numbers (8.NS.A.1)

1. Which of the following is an integer?

   Ⓐ -3
   Ⓑ $\frac{1}{4}$
   Ⓒ -12.5
   Ⓓ 0.454545...

2. Which of the following statements is true?

   Ⓐ Every rational number is an integer.
   Ⓑ Every whole number is a rational number.
   Ⓒ Every irrational number is a natural number.
   Ⓓ Every rational number is a whole number.

3. Which of the following most accurately describes the square root of 10?

   Ⓐ It is rational.
   Ⓑ It is irrational.
   Ⓒ It is an integer.
   Ⓓ It is a whole number.

4. Complete the following statement: Pi is _____ .

   Ⓐ both real and rational
   Ⓑ real but not rational
   Ⓒ rational but not real
   Ⓓ neither real nor rational

5. Complete the following statement: $\sqrt{7}$ is _____.

   Ⓐ both a real and a rational number
   Ⓑ a real number, but not rational
   Ⓒ a rational number, but not a real number
   Ⓓ neither a real nor a rational number

6. The number 57 belongs to which of the following set(s) of numbers?

    Ⓐ N only
    Ⓑ N, W, and Z only
    Ⓒ N, W, Z, and Q only
    Ⓓ All of the following: N, W, Z, Q, and R

7. From the following set: {-√5.7, -9, 0, 5.25, 3i, √16}
    Select the answer choice that shows the elements which are Natural numbers.

    Ⓐ -√5.7, -9, 0, 5.25, 3i, √16
    Ⓑ -√5.7, -9, 0, 5.25, 3i
    Ⓒ 3i
    Ⓓ Positive square root of 16

8. From the following set: {-√5.7, -9, 0, 5.25, 3i, √16}
    Select the answer choice that shows the elements that are Rational numbers.

    Ⓐ -√5.7, -9, 0, 5.25, 3i, √16
    Ⓑ -9, 0, 5.25, √16
    Ⓒ 3i
    Ⓓ -√5.7

9. Which of the numbers below is irrational?

    Ⓐ √169
    Ⓑ √4
    Ⓒ √16
    Ⓓ √3

10. Write the repeating rational number 0.1515... as a fraction.

    Ⓐ $\frac{85}{100}$

    Ⓑ $\frac{15}{75}$

    Ⓒ $\frac{15}{99}$

    Ⓓ $\frac{25}{50}$

11. Write the repeating rational number .112112... as a fraction.

Ⓐ $\dfrac{112}{100}$

Ⓑ $\dfrac{112}{99}$

Ⓒ $\dfrac{112}{999}$

Ⓓ $\dfrac{111}{999}$

12. Which of the following is true of the square root of 2?

Ⓐ It is both real and rational.
Ⓑ It is real but not rational.
Ⓒ It is rational but not real.
Ⓓ It is neither real nor rational.

13. Which of the following sets includes the square root of -25?

Ⓐ R, Z, W, N, and Q
Ⓑ R, W, Q
Ⓒ Z, N
Ⓓ None of the above.

14. Complete the following statement:
The number 6.25 belongs to _____.

Ⓐ R, Q, Z, W, and N
Ⓑ R and Q
Ⓒ R and N
Ⓓ Q and Z

15. Complete the sentence:
Irrational numbers may always be written as _____.

Ⓐ fractions
Ⓑ fractions and as decimals
Ⓒ decimals but not as fractions
Ⓓ neither decimals or fractions

Approximating Irrational Numbers (8.NS.A.2)

1. Between which two whole numbers does √5 lie on the number line?

    Ⓐ 1 and 2
    Ⓑ 2 and 3
    Ⓒ 3 and 4
    Ⓓ 4 and 5

2. Between which pairs of rational numbers does √5 lie on the number line?

    Ⓐ 2.0 and 2.1
    Ⓑ 2.1 and 2.2
    Ⓒ 2.2 and 2.3
    Ⓓ 2.3 and 2.4

3. Order the following numbers on a number line (least to greatest).

    Ⓐ 1.8, 1.35, 2,5, √5
    Ⓑ 1.35,√5, 1.8, 2.5
    Ⓒ 1.35, 1.8, √5, 2.5
    Ⓓ 1.35, 1.8, 2.5, √5

4. If you fill in the _____ in each of the following choices with √7, which displays the correct ordering from least to greatest?

    Ⓐ ___, 2.5, 2.63, 2.65
    Ⓑ 2.5, ___, 2.63, 2.65
    Ⓒ 2.5, 2.63, ___, 2.65
    Ⓓ 2.5, 2.63, 2.65, ___

5. Which of the following numbers has the least value?

    Ⓐ √(0.6561)
    Ⓑ .8
    Ⓒ .8...
    Ⓓ .8884

6. **Choose the correct order (least to greatest) for the following real numbers.**

Ⓐ $\sqrt{5}$, $4\frac{1}{2}$, 4.75, $2\sqrt{10}$

Ⓑ $4\frac{1}{2}$, $\sqrt{5}$, $2\sqrt{10}$, 4.75

Ⓒ $4\frac{1}{2}$, 4.75, $\sqrt{5}$, $2\sqrt{10}$

Ⓓ $\sqrt{5}$, $2\sqrt{10}$, $4\frac{1}{2}$, 4.75

7. **Which of the following numbers has the greatest value?**

Ⓐ .4...
Ⓑ .444
Ⓒ $\sqrt{.4}$
Ⓓ .45

8. **Which is the correct order of the following numbers when numbering from least to greatest?**

Ⓐ $\sqrt{.9}$, .9, .999, .9...
Ⓑ .9, $\sqrt{.9}$, .999, .9...
Ⓒ .9, .9..., $\sqrt{.9}$, .999
Ⓓ .9, .9..., .999, $\sqrt{.9}$

9. **Write the following numbers from least to greatest.**

Ⓐ $\sqrt{2}$, π, $3\frac{7}{8}$, $\frac{32}{8}$

Ⓑ π, $\sqrt{2}$, $3\frac{7}{8}$, $\frac{32}{8}$

Ⓒ $3\frac{7}{8}$, π, $\sqrt{2}$, $\frac{32}{8}$

Ⓓ $\frac{32}{8}$, $3\frac{7}{8}$, π, $\sqrt{2}$

10. If you were to arrange the following numbers on the number line from least to greatest, which one would be last?

Ⓐ 3.6

Ⓑ $3\frac{7}{12}$

Ⓒ $\sqrt{12}$

Ⓓ $3\frac{9}{10}$

11. Between which of these pairs of rational numbers does $\sqrt{24}$ lie on the number line?

Ⓐ 4.79 and $4\frac{7}{8}$

Ⓑ $4\frac{7}{8}$ and 5.0

Ⓒ 4.95 and 5.0

Ⓓ 4.75 and 4.79

12. Between which two integers does $\sqrt{2}$ lie on the number line?

Ⓐ 0 and 1
Ⓑ 1 and 2
Ⓒ 2 and 3
Ⓓ 3 and 4

13. Between which pair of rational numbers does $\sqrt{2}$ lie on the number line?

Ⓐ 1.40 and 1.41
Ⓑ 1.41 and 1.42
Ⓒ 1.42 and 1.43
Ⓓ 1.43 and 1.44

14. Between which pair of consecutive integers on the number line does $\sqrt{3}$ lie?

Ⓐ 1 and 2
Ⓑ 2 and 3
Ⓒ 3 and 4
Ⓓ 4 and 5

15. **Between which of the following pairs of rational numbers on the number line does √3 lie?**

    Ⓐ 1.70 and 1.71
    Ⓑ 1.71 and 1.72
    Ⓒ 1.72 and 1.73
    Ⓓ 1.73 and 1.74

End of The Number System

Answer Key and Detailed Explanations
The Number System
Rational vs. Irrational Numbers (8.NS.A.1)

Question No.	Answer	Detailed Explanation
1	A	An integer belongs to the set containing the counting numbers, their additive inverses, and zero. Therefore, (-3) is an integer.
2	B	Rational numbers are the set of numbers that can be expressed as the quotient of two integers in which the denominator is not zero. All whole numbers can be expressed in this manner; so every whole number is a rational number.
3	B	√10 cannot be expressed as the ratio of two integers p and q and is therefore irrational.
4	B	Pi is the ratio of a circle's circumference to its diameter. It is therefore a real number. Pi cannot be expressed as the ratio of two integers, so it is irrational.
5	B	√7 cannot be expressed as the ratio of two integers and is therefore irrational. The irrationals are a subset of the real numbers.
6	D	The number 57 meets the requirements of each of the following sets of numbers: N (natural numbers), W (whole numbers), Z (integers), Q (rational numbers), and R (real numbers).
7	D	By definition, the natural numbers, N, are the set of counting numbers. Some mathematicians also include zero in this set. Since √16 = +4 or -4 and +4 is a counting number, it is included in N. None of the choices offered 0 as an option; so, in this case, it is a mute point.
8	B	3i is an imaginary number and therefore not rational and - √5.7 cannot be expressed as a terminating or repeating decimal and consequently is not rational. Therefore, there is only one choice that does not include one or the other or both of these two options. Option B is the correct answer.
9	D	√3 is non-terminating and non-repeating.

Question No.	Answer	Detailed Explanation
10	C	1- Write equation 1 - Assign the repeating rational number to x: x = 0.1515... 2- Write equation 2 - Multiply equation 1 by 100: 100x = 15.1515... 3- Subtract equation 1 from 2: 99x = 15 x = 15/99
11	C	1- Write equation 1 - Assign the repeating rational number to x: x = .112112... 2- Write equation 2 - Multiply equation 1 by 1000: 1000x = 112.112... 3- Subtract equation 1 from 2: 1000x = 112.112 x = 112/999
12	B	$\sqrt{2}$ cannot be expressed as a non-terminating or non-repeating decimal and is therefore irrational. However, it is real.
13	D	The square root of a negative number is not a real number. Since the first three choices contain real numbers, none of them will fit.
14	B	Numbers with terminating decimals are real (R) and rational (Q), but are not integers (Z), whole numbers (W), or natural numbers (N).
15	C	By definition all rational numbers may be written as terminating or repeating decimals. Irrational numbers can be written as decimals (non-repeating, non-terminating), but not as fractions.

Approximating Irrational Numbers (8.NS.A.2)

1	B	$2^2=4$ and $3^2=9$ Since 5 lies between 4 and 9, $\sqrt{5}$ lies between 2 and 3.
2	C	$(2.2)^2=4.84$ and $(2.3)^2=5.29$ Since 5 lies between 4.84 and 5.29 on the number line, $\sqrt{5}$ lies between 2.2 and 2.3.
3	C	If we change the numbers all to the same accuracy, it is easier to order them on the number line. Write 1.8 as 1.80, 2.5 as 2.50 and $\sqrt{5}$ as >2 because $2^2=4$ and < 2.5 because $2.5^2=6.25$. Then, the correct order is 1.35, 1.8, $\sqrt{5}$, 2.5.

Question No.	Answer	Detailed Explanation
4	C	$2.63^2 \approx 6.92$ $2.65^2 \approx 7.02$ The square root of 7 is about 2.64, so $\sqrt{7}$ falls between 2.63 and 2.65. Then, 2.5, 2.63, $\sqrt{7}$, 2.65 is the correct answer.
5	B	$\sqrt{(0.6561)} = .81$ .8... means that the 8 is repeating; i.e. .888... .8 may be written as .80 So .8 represents the smallest (least) value.
6	A	$4\ 1/2 = 4.5$ $2 < \sqrt{5} < 3$ $\sqrt{10} > 3$; so $2\sqrt{10} > 6$ Then, the correct order is: $\sqrt{5}$, 4 1/2, 4.75, $2\sqrt{10}$
7	C	$\sqrt{.4} = $ (approx.) .63 which is the greatest of these numbers.
8	B	$\sqrt{.9} = $ (approx.) .95 So $\sqrt{.9}$, .9, .999, .9... is the correct order.
9	A	$32/8 = 4$, $\pi = $ (approx) 3.14, $\sqrt{2} = $ (approx) 1.4 The correct order is $\sqrt{2}$, π, 3 7/8, 32/8
10	D	Convert each of the numbers to a similar expression. $\sqrt{12} = $ approx. 3.46 $3\ 7/12 = $ approx. 3.58 $3\ 9/10 = 3.9$ The correct order is $\sqrt{12}$, 3 7/12, 3.6, 3 9/10
11	B	$\sqrt{24} = $ approx. 4.90 $4\ 7/8 = 4.875$ So 4 7/8 and 5.0 is the correct answer.
12	B	$1^2 = 1$ and $2^2 = 4$ Since 2 >1 and <4, $\sqrt{2}$ lies between 1 and 2.
13	B	$1.41^2 = 1.974$ and $1.42^2 = 2.0164$ Since 2 lies between 1.974 and 2.0164 on the number line, $\sqrt{2}$ lies between 1.41 and 1.42 on the number line.
14	A	$1^2 = 1$ and $2^2 = 4$ Since 3 lies between 1^2 and 2^2, Then $\sqrt{3}$ lies between 1 and 2.
15	D	$1.73^2 = 2.9929$ and $1.74^2 = 3.0276$ Since 3 lies between 2.9929 and 3.0276 on the number line, $\sqrt{3}$ lies between 1.73 and 1.74.

Expressions and Equations

Properties of Exponents (8.EE.A.1)

1. Is -5^2 equal to $(-5)^2$?

   Ⓐ Yes, because they both equal -25.
   Ⓑ Yes, because they both equal -10.
   Ⓒ Yes, because they both equal 25.
   Ⓓ No, because -5^2 equals -25 and $(-5)^2$ equals 25.

2. $\dfrac{X^6}{X^{-2}} =$

   Ⓐ $\dfrac{1}{X^3}$

   Ⓑ $\dfrac{1}{X^{12}}$

   Ⓒ X^4

   Ⓓ X^8

3. Which of the following is equal to 3^{-2} ?

   Ⓐ $\dfrac{1}{9}$

   Ⓑ $\dfrac{1}{6}$

   Ⓒ -9

   Ⓓ 9

4. Which of the following is equivalent to $X^{(2-5)}$?

Ⓐ X^3

Ⓑ $X^{\frac{1}{3}}$

Ⓒ $\dfrac{1}{X^3}$

Ⓓ 3^X

5. $1^9 =$

Ⓐ 1
Ⓑ 3
Ⓒ 9
Ⓓ $\dfrac{1}{9}$

6. $(X^{-3})(X^{-3}) =$

Ⓐ X^6
Ⓑ X^9
Ⓒ $\dfrac{1}{X^6}$
Ⓓ $\dfrac{1}{X^9}$

7. $(X^{-2})^{-7} =$

Ⓐ X^5

Ⓑ X^{14}

Ⓒ $\dfrac{1}{X^5}$

Ⓓ $\dfrac{1}{X^{14}}$

8. $(X^4)^0 =$

Ⓐ X
Ⓑ X^4
Ⓒ 1
Ⓓ 0

9. $(3^2)^3 =$

   Ⓐ 3^5
   Ⓑ 3^6
   Ⓒ 3
   Ⓓ 1

10. $5^2 + 5^3 =$

   Ⓐ 150
   Ⓑ 125
   Ⓒ 25
   Ⓓ 10

Square & Cube Roots (8.EE.A.2)

1. What is the cube root of 1,000 ?

   Ⓐ 10

   Ⓑ 100

   Ⓒ $33\frac{1}{3}$

   Ⓓ $333\frac{1}{3}$

2. $8\sqrt{12} \div \sqrt{15} =$

   Ⓐ $\frac{4}{5}$

   Ⓑ $\frac{8}{5}$

   Ⓒ $\frac{16}{\sqrt{5}}$

   Ⓓ $\frac{\sqrt{5}}{8}$

3. The square root of 75 is between which two integers?

   Ⓐ 8 and 9
   Ⓑ 7 and 8
   Ⓒ 9 and 10
   Ⓓ 6 and 7

4. The square root of 110 is between which two integers?

   Ⓐ 10 and 11
   Ⓑ 9 and 10
   Ⓒ 11 and 12
   Ⓓ 8 and 9

5. Solve the following problem: $6\sqrt{20} \div \sqrt{5} =$ _____

   Ⓐ 12
   Ⓑ 11
   Ⓒ 30
   Ⓓ 5

6. The cube root of 66 is between which two integers?

   Ⓐ 4 and 5
   Ⓑ 3 and 4
   Ⓒ 5 and 6
   Ⓓ 6 and 7

7. Which expression has the same value as $3\sqrt{144} \div \sqrt{12}$?

   Ⓐ $36 \div \sqrt{12}$
   Ⓑ $3\sqrt{12}$
   Ⓒ $27 \div \sqrt{12}$
   Ⓓ $33 \div \sqrt{12}$

8. The cubic root of 400 lies between which two numbers?

   Ⓐ 5 and 6
   Ⓑ 6 and 7
   Ⓒ 7 and 8
   Ⓓ 8 and 9

9. Which of the following is equivalent to the expression $4\sqrt{250} \div 5\sqrt{2}$?

   Ⓐ $4\sqrt{25} \div 5$
   Ⓑ $4\sqrt{125} \div \sqrt{2}$
   Ⓒ $4\sqrt{10}$
   Ⓓ $4\sqrt{5}$

10. The cube root of 150 is closest to which of the following?

   Ⓐ 15
   Ⓑ 10
   Ⓒ 5
   Ⓓ 3

Scientific Notation (8.EE.A.3)

1. In 2007, approximately 3,380,000 people visited the Statue of Liberty. Express this number in scientific notation.

   Ⓐ 0.388×10^7
   Ⓑ 3.38×10^6
   Ⓒ 33.8×10^5
   Ⓓ 338×10^1

2. The average distance from Saturn to the Sun is 890,800,000 miles. Express this number in scientific notation.

   Ⓐ 8908×10^8
   Ⓑ 8908×10^5
   Ⓒ 8.908×10^8
   Ⓓ 8.908×10^5

3. The approximate population of Los Angeles is 3.8×10^6 people. Express this number in standard notation.

   Ⓐ 380,000
   Ⓑ 3,800,000
   Ⓒ 38,000,000
   Ⓓ 380,000,000

4. The approximate population of Kazakhstan is 1.53×10^7 people. Express this number in standard notation.

   Ⓐ 153,000
   Ⓑ 1,530,000
   Ⓒ 15,300,000
   Ⓓ 153,000,000

5. The typical human body contains about 2.5×10^{-3} kilograms of zinc. Express this amount in standard form.

   Ⓐ 0.00025 kilograms
   Ⓑ 0.0025 kilograms
   Ⓒ 0.025 kilograms
   Ⓓ 0.25 kilograms

6. If a number expressed in scientific notation is N x 10^5, how large is the number?

Ⓐ Between 1,000 (included) and 10,000
Ⓑ Between 10,000 (included) and 100,000
Ⓒ Between 100,000 (included) and 1,000,000
Ⓓ Between 1,000,000 (included) and 10,000,000

7. Red light has a wavelength of 650 x 10^{-9} meters. Express the wavelength in scientific notation.

Ⓐ 65.0 x 10^{-10} meters ✗
Ⓑ 65.0 x 10^{-8} meters
Ⓒ 6.50 x 10^{-7} meters
Ⓓ 6.50 x 10^{-11} meters ✗

8. A strand of hair from a human head is approximately 1 x 10^{-4} meters thick. What fraction of a meter is this?

Ⓐ $\dfrac{1}{100}$

Ⓑ $\dfrac{1}{1,000}$

Ⓒ $\dfrac{1}{10,000}$

Ⓓ $\dfrac{1}{100,000}$

9. Which of the following numbers has the greatest value?

Ⓐ 8.93 x 10^3
Ⓑ 8.935 x 10^2
Ⓒ 8.935 x 10^3
Ⓓ 893.5 x 10^1

10. Which of the following numbers has the least value?

Ⓐ -1.56 x 10^2
Ⓑ -1.56 x 10^3
Ⓒ 1.56 x 10^2
Ⓓ 1.56 x 10^3

Solving Problems Involving Scientific Notation (8.EE.A.4)

1. The population of California is approximately 3.7×10^7 people. The land area of California is approximately 1.6×10^5 square miles. Divide the population by the area to find the best estimate of the number of people per square mile in California.

    Ⓐ 24 people
    Ⓑ 240 people
    Ⓒ 2,400 people
    Ⓓ 24,000 people

2. Mercury is approximately 6×10^7 kilometers from the Sun. The speed of light is approximately 3×10^5 kilometers per second. Divide the distance by the speed of light to determine the approximate number of seconds it takes light to travel from the Sun to Mercury.

    Ⓐ 2 seconds
    Ⓑ 20 seconds
    Ⓒ 200 seconds
    Ⓓ 2,000 seconds

3. Simplify $(4 \times 10^6) \times (2 \times 10^3)$ and express the result in scientific notation.

    Ⓐ 8×10^9
    Ⓑ 8×10^{18}
    Ⓒ 6×10^9
    Ⓓ 6×10^{18}

4. Simplify $(2 \times 10^{-3}) \times (3 \times 10^5)$ and express the result in scientific notation.

    Ⓐ 5×10^{-8}
    Ⓑ 5×10^{-15}
    Ⓒ 6×10^8
    Ⓓ 6×10^2

5. Washington is approximately 2.4×10^3 miles from Utah. Mary drives 6×10 miles per hour from Washington to Utah. Divide the distance by the speed to determine the approximate number of hours it takes Mary to travel from Washington to Utah.

    Ⓐ 41 hours
    Ⓑ 40 hours
    Ⓒ 39 hours
    Ⓓ 38 hours

6. Which of the following is NOT equal to $(5 \times 10^5) \times (9 \times 10^{-3})$?

   Ⓐ 4.5×10^4
   Ⓑ 4.5×10^3
   Ⓒ $4,500$
   Ⓓ 45×100

7. Find $(5 \times 10^7) \div (10 \times 10^2)$ and express the result in scientific notation.

   Ⓐ 5×10^4
   Ⓑ 0.5×10^5
   Ⓒ 50×10^9
   Ⓓ 5.0×10^9

8. Approximate $.00004567 \times .00001234$ and express the result in scientific notation.

   Ⓐ 5.636
   Ⓑ 5.636×10^0
   Ⓒ 5.636×10^{-10}
   Ⓓ None of the above.

9. Find the product $(50.67 \times 10^4) \times (12.9 \times 10^3)$ and express the answer in standard notation.

   Ⓐ 653.643
   Ⓑ 653.643×10^7
   Ⓒ 6.53643×10^9
   Ⓓ $6,536,000,000$

10 Approximate the quotient and express the answer in standard notation.
   $(1.298 \times 10^4) \div (3.97 \times 10^2)$

   Ⓐ 32.7
   Ⓑ $.327$
   Ⓒ $.327 \times 10^2$
   Ⓓ None of the above.

Compare Proportions (8.EE.B.5)

1. Find the unit rate if 12 tablet cost $1,440.

   Ⓐ $100
   Ⓑ $150
   Ⓒ $120
   Ⓓ $50

2. A package of Big Bubbles Gum has 10 pieces and sells for $2.90. A package of Fruity Gum has 20 pieces and sells for $6.20. Compare the unit prices.

   Ⓐ Big Bubbles is $0.10 more per piece than Fruity.
   Ⓑ Fruity is $0.02 more per piece than Big Bubbles.
   Ⓒ They both have the same unit price.
   Ⓓ It cannot be determined.

3. The first major ski slope in Vermont has a rise of 9 feet vertically for every 54 feet horizontally. A second ski slope has a rise of 12 feet vertically for every 84 feet horizotally. Which of the following statements is true?

   Ⓐ The first slope is steeper than the second.
   Ⓑ The second slope is steeper than the first.
   Ⓒ Both slopes have the same steepness.
   Ⓓ Cannot be determined from the information given.

4. Which of the following ramps has the steepest slope?

   Ⓐ Ramp A has a vertical rise of 3 feet and a horizontal run of 15 feet
   Ⓑ Ramp B has a vertical rise of 4 feet and a horizontal run of 12 feet
   Ⓒ Ramp C has a vertical rise of 2 feet and a horizontal run of 10 feet
   Ⓓ Ramp D has a vertical rise of 5 feet and a horizontal run of 20 feet

5. Choose the statement that is true about unit rate.

   Ⓐ The unit rate can also be called the rate of change.
   Ⓑ The unit rate can also be called the mode.
   Ⓒ The unit rate can also be called the frequency.
   Ⓓ The unit rate can also be called the median.

6. Which statement is false?

   Ⓐ Unit cost is calculated by dividing the amount of items by the total cost.
   Ⓑ Unit cost is calculated by dividing the total cost by the amount of items.
   Ⓒ Unit cost is the cost of one unit item.
   Ⓓ On similar items, a higher unit cost is not the better price.

7. Selena is preparing for her eighth grade graduation party. She must keep within the budget set by her parents. Which is the best price for her to purchase ice cream?

   Ⓐ $3.99/ 24 oz carton
   Ⓑ $4.80/ one-quart carton
   Ⓒ $11.00 / one gallon tub
   Ⓓ $49.60/ five gallon tub

8. Ben is building a ramp for his skate boarding club. Which of the following provides the least steep ramp?

   Ⓐ 2 feet vertical for every 10 feet horizontal
   Ⓑ 3 feet vertical for every 9 feet horizontal
   Ⓒ 4 feet vertical for every 16 feet horizontal
   Ⓓ 5 feet vertical for every 30 feet horizontal

9. Riley is shopping for tee shirts. Which is the most expensive (based on unit price per shirt)?

   Ⓐ 5 tee shirts for $50.00
   Ⓑ 6 tee shirts for $90.00
   Ⓒ 2 tee shirts for $22.00
   Ⓓ 4 tee shirts for $48.00

10. The swim team is preparing for a meet. Which of the following is Lindy's fastest time?

    Ⓐ five laps in fifteen minutes
    Ⓑ four laps in sixteen minutes
    Ⓒ two laps in ten minutes
    Ⓓ three laps in eighteen minutes

11. David is having a Super Bowl party and he needs bottled sodas. Which of the following purchases will give him the lowest unit cost?

Ⓐ $2.00 for a 6 pack
Ⓑ $6.00 for a 24 pack
Ⓒ $3.60 for a 12 pack
Ⓓ $10.00 for a 36 pack

12. A package of plain wafers has 20 per pack and sells for $2.40. A package of sugar-free wafers has 30 pieces and sells for $6.30. Compare the unit prices.

Ⓐ Each plain wafer is $0.17 more than a sugar-free wafer.
Ⓑ Each sugar-free wafer is $0.09 more than a plain wafer.
Ⓒ They both have the same unit price per wafer.
Ⓓ The relationship cannot be determined.

13. Li took 4 practice tests to prepare for his chapter test.
    Which of the following is the best score?

Ⓐ 36 correct out of 40 questions
Ⓑ 24 correct out of 30 questions
Ⓒ 17 correct out of 25 questions
Ⓓ 15 correct out of 20 questions

14. Mel's class is planning a fundraiser. They have decided to have a carnival. If they sell tickets in packs of 40 for $30.00, what is the unit cost?

Ⓐ $4.00 per ticket
Ⓑ $0.50 per ticket
Ⓒ $1.75 per ticket
Ⓓ $0.75 per ticket

15. Which of the following ski slopes has the steepest slope?

Ⓐ Ski Slope A has a vertical rise of 4 feet and a horizontal run of 16 feet
Ⓑ Ski Slope B has a vertical rise of 3 feet and a horizontal run of 12 feet
Ⓒ Ski Slope C has a vertical rise of 3 feet and a horizontal run of 9 feet
Ⓓ Ski Slope D has a vertical rise of 5 feet and a horizontal run of 25 feet

Name: _____   Date: _____

1. **Which of the following statements is true about slope?**

    Ⓐ Slopes of straight lines will always be positive numbers.
    Ⓑ The slopes vary between the points on a straight line.
    Ⓒ Slope is determined by dividing the horizontal distance between two points by the corresponding vertical distance.
    Ⓓ Slope is determined by dividing the vertical distance between two points by the corresponding horizontal distance.

2. **Which of the following is an equation of the line passing through the points (-1, 4) and (1, 2)?**

    Ⓐ $y = x - 3$
    Ⓑ $y = 2x + 2$
    Ⓒ $y = -2x + 4$
    Ⓓ $y = -x + 3$

3. **The graph of which equation has the same slope as the graph of $y = 4x + 3$?**

    Ⓐ $y = -2x + 3$
    Ⓑ $y = 2x - 3$
    Ⓒ $y = -4x + 2$
    Ⓓ $y = 4x - 2$

4. **Which of these lines has the greatest slope?**

    Ⓐ $y = \frac{8}{5}x - 7$

    Ⓑ $y = \frac{6}{5}x + 4$

    Ⓒ $y = \frac{7}{5}x + 2$

    Ⓓ $y = \frac{9}{5}x - 3$

5. **Which of these lines has the smallest slope?**

    Ⓐ $y - \frac{1}{8}x + 7$

    Ⓑ $y - \frac{1}{3}x + 7$

    Ⓒ $y - \frac{1}{4}x - 9$

    Ⓓ $y = \frac{1}{7}x$

6. Fill in the blank with one of the four choices to make the following a true statement. Knowing _____ and the y-intercept is NOT enough for us to write the equation of the line.

    Ⓐ direction
    Ⓑ a point on a given line
    Ⓒ the x-intercept
    Ⓓ the slope

7. A skateboarder is practicing at the city park. He is skating up and down the steepest straight line ramp. If the highest point on the ramp is 30 feet above the ground and the horizontal distance from the base of the ramp to a point directly beneath the upper end is 500 feet, what is the slope of the ramp?

    Ⓐ $\frac{500}{30}$

    Ⓑ $\frac{50}{3}$

    Ⓒ $\frac{3}{50}$

    Ⓓ None of these.

8. If the equation of a line is expressed as $y = \frac{3}{2}x - 9$, what is the slope of the line?

    Ⓐ - 9

    Ⓑ +9

    Ⓒ $\frac{3}{2}$

    Ⓓ $\frac{2}{3}$

9. Which of the following is an equation of the line that passes through the points (0, 5) and (2, 15)?

    Ⓐ y = 5x + 5
    Ⓑ y = 5x + 3
    Ⓒ y = 3x + 5
    Ⓓ y = 5x - 5

10. Which of the following equations has the same slope as the line passing through the points (1, 6) and (3, 10)?

Ⓐ y = 2x - 9
Ⓑ y = 5x - 2
Ⓒ y = 4x - 5
Ⓓ y = 9x - 6

11. Which of the following equations has the same slope as the line passing through the points (3, 6) and (5, 10)?

Ⓐ y = 2x -12
Ⓑ y = 11x - 8
Ⓒ y = -2x - 9
Ⓓ y = 3x - 5

12. Which equation has the same slope as y = -5x - 4?

Ⓐ y = 5x +15
Ⓑ y = -5x - 11
Ⓒ y = 5x -19
Ⓓ y = 5x - 13

13. Find the slope of the line passing through the points (3,3) and (5,5).

Ⓐ 2
Ⓑ 1
Ⓒ 3
Ⓓ 5

14. Which of the following lines has the steepest slope?

Ⓐ y = 4x+5
Ⓑ y =-3x + 5
Ⓒ y = 3x - 5
Ⓓ They all have the same slope.

15. Which of the following is the equation of the line passing through the points (0,-3) and (-3,0) ?

Ⓐ y = -3x + 3
Ⓑ y = -3x
Ⓒ y = -x - 3
Ⓓ y = -x

Solving Linear Equations (8.EE.C.7.A)

1. Which two consecutive odd integers have a sum of 44?

   Ⓐ 21 and 23
   Ⓑ 19 and 21
   Ⓒ 23 and 25
   Ⓓ 17 and 19

2. During each of the first three quarters of the school year, Melissa earned a grade point average of 2.1, 2.9, and 3.1. What does her 4th quarter grade point average need to be in order to raise her grade to a 3.0 cumulative grade point average?

   Ⓐ 3.9
   Ⓑ 4.2
   Ⓒ 2.6
   Ⓓ 3.5

3. Martha is on a trip of 1,924 miles. She has already traveled 490 miles. She has 3 days left in her trip. How many miles does she need to travel each day to complete her trip?

   Ⓐ 450 miles/day
   Ⓑ 464 miles/day
   Ⓒ 478 miles/day
   Ⓓ 492 miles/day

4. Find the solution to the following equation: $3x + 5 = 29$

   Ⓐ $x = 24$
   Ⓑ $x = 11$
   Ⓒ $x = 8$
   Ⓓ $x = 6$

5. Find the solution to the following equation:
   $7 - 2x = 13 - 2x$

   Ⓐ $x = -10$
   Ⓑ $x = -3$
   Ⓒ $x = 3$
   Ⓓ There is no solution.

6. Find the solution to the following equation: 6x + 1 = 4x - 3

    Ⓐ x = -1
    Ⓑ x = -2
    Ⓒ x = - 0.5
    Ⓓ There is no solution.

7. Find the solution to the following equation:
   2x + 6 + 1 = 7 + 2x

    Ⓐ x = -3
    Ⓑ x = 3
    Ⓒ x = 7
    Ⓓ All real numbers are solutions.

8. Find the solution to the following equation: 8x-1=8x

    Ⓐ x = 7
    Ⓑ x = -8
    Ⓒ x = 8
    Ⓓ There is no solution.

9. Which of the answers is the correct solution to the following equation?
   2x + 5x - 9 = 8x - x - 3 - 6

    Ⓐ x = 3
    Ⓑ x = 7
    Ⓒ x = 9
    Ⓓ All real values for x are correct solutions.

10. Solve the following linear equation for y.
   -y + 7y -54 = 0

    Ⓐ y = 0
    Ⓑ y = 1
    Ⓒ y = 6
    Ⓓ y = 9

Solve Linear Equations with Rational Numbers (8.EE.C.7.B)

1. Solve the following linear equation: $\frac{7}{14} = n + \frac{7}{14}n$

   Ⓐ $n = 1\frac{1}{2}$

   Ⓑ $n = 3$

   Ⓒ $n = \frac{1}{3}$

   Ⓓ $n = 1$

2. Find the solution to the following equation: $2(2x - 7) = 14$

   Ⓐ $x = 14$
   Ⓑ $x = 7$
   Ⓒ $x = 1$
   Ⓓ $x = 0$

3. Solve the following equation for x.
   $6x - (2x + 5) = 11$

   Ⓐ $x = -3$
   Ⓑ $x = -4$
   Ⓒ $x = 3$
   Ⓓ $x = 4$

4. $4x + 2(x - 3) = 0$

   Ⓐ $x = 0$
   Ⓑ $x = 1$
   Ⓒ $x = 2$
   Ⓓ All real values for x are correct solutions.

5. Solve the following equation for y.
   $3y - 7(y + 5) = y - 35$

   Ⓐ $y = 0$
   Ⓑ $y = 1$
   Ⓒ $y = 2$
   Ⓓ All real values for y are correct solutions.

6. Solve the following linear equation: $2(x-5) = \frac{1}{2}(6x+4)$

   Ⓐ x= -12
   Ⓑ x= -9
   Ⓒ x= -4
   Ⓓ There is no solution.

7. Solve the following linear equation for x.

   $3x + 2 + x = \frac{1}{3}(12x + 6)$

   Ⓐ x= -4
   Ⓑ x= 2
   Ⓒ There is no solution.
   Ⓓ All real values for x are correct solutions.

8. $\frac{1}{2}x + \frac{2}{3}x + 5 = \frac{5}{2}x + 6$

   Ⓐ x = $\frac{33}{4}$

   Ⓑ x = $\frac{1}{2}$

   Ⓒ x = $-\frac{3}{4}$

   Ⓓ x = $-\frac{6}{5}$

9. Which of the following could be a correct procedure for solving the equation below?
   $2(2x+3) = 3(2x+5)$

   Ⓐ 4x+5 = 6x+5
       -2x+5 = 5
       -2x = 0
       x = 0

   Ⓒ 2(5x) = 6x+15
       10x = 6x+15
       4x = 15

       x = $\frac{15}{4}$

   Ⓑ 4x+6 = 6x+15
       -2x+6 = 15
       -2x = 9

       x = $-\frac{9}{2}$

   Ⓓ 4x+6 = 6x+15
       -2x+6 = 15
       -2x = 9

       x = $\frac{2}{9}$

10. Solve the following linear equation:
   0.64x - 0.15x + 0.08 = 0.09x

   Ⓐ x = - 5
   Ⓑ x = - 0.2
   Ⓒ x = 5.125
   Ⓓ There is no solution.

Solutions to Systems of Equations (8.EE.C.8.A)

1. Which of the following points is the intersection of the graphs of the lines given by the equations y = x - 5 and y = 2x + 1 ?

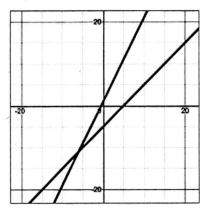

   Ⓐ (1, 3)
   Ⓑ (-1, -4)
   Ⓒ (-2, -3)
   Ⓓ (-6, -11)

2. Which of the following best describes the solution set of this system?
   y = 0.5x + 7
   y = 0.5x - 1

   Ⓐ The solution is (-2, -3) because the graphs of the two equations intersect at that point.
   Ⓑ The solution is (0.5, 3) because the graphs of the two equations intersect at that point.
   Ⓒ There is no solution because the graphs of the two equations are parallel lines.
   Ⓓ There are infinitely many solutions because the graphs of the two equations are the same line.

3. Find the solution to the following system:
   y = 2(2 - 3x)
   y = -3(2x + 3)

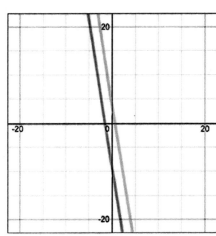

   Ⓐ x = -1; y = 10
   Ⓑ x = -2; y = 24
   Ⓒ x = -3; y = 22
   Ⓓ There is no solution.

4. Use the graph, to find the solution to the following system:

$\frac{x}{2} + \frac{y}{3} = 2$

$3x - 2y = 48$

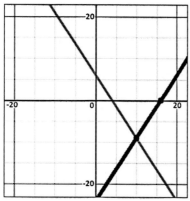

Ⓐ x = 8, y = -6
Ⓑ x = 10, y = -9
Ⓒ x = 12, y = -3
Ⓓ x = 16, y = 0

5. Which of the following best describes the relationship between the graphs of the equations in this system?
$y = 2x - 6$
$y = -2x + 6$

Ⓐ The lines intersect at the point (0, -3).
Ⓑ The lines intersect at the point (3, 0).
Ⓒ The lines do not intersect because their slopes are opposites and their y-intercepts are opposites.
Ⓓ They are the same line because their slopes are opposites and their y-intercepts are opposites.

6.   Which graph shows the solution to the system y = -x + 3 and y = -2x + 8?

Ⓐ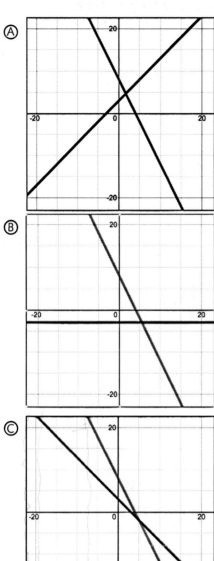

Ⓑ

Ⓒ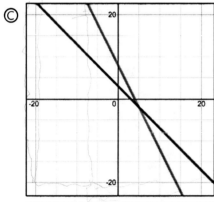

Ⓓ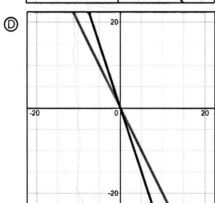

7. **Which graph shows the solution to the system** $2x+y = 12$ **and** $y = \dfrac{1}{5}x$**?**

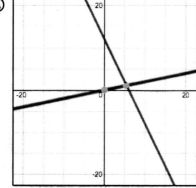

Ⓐ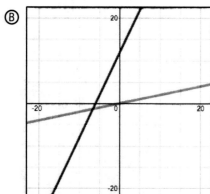

Ⓑ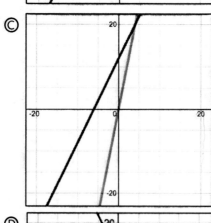

Ⓒ

Ⓓ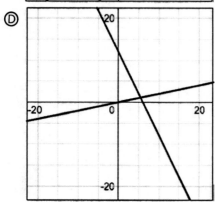

8. Which graph represents a system of equations with infinite solutions?

Ⓐ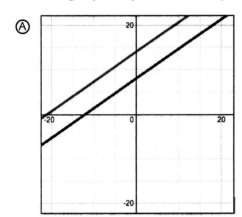

Ⓑ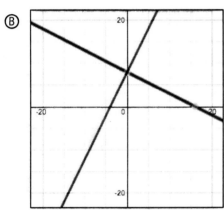

Ⓒ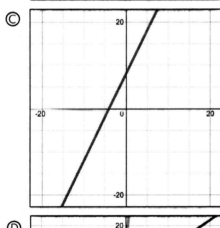

Ⓓ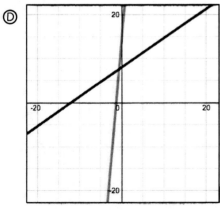

9. **Which graph represents a system of equations with no solutions?**

Ⓐ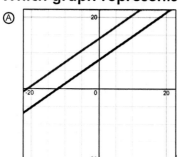

Ⓑ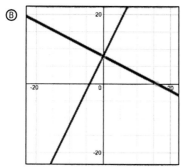

Ⓒ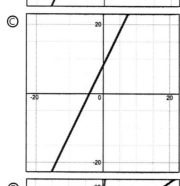

Ⓓ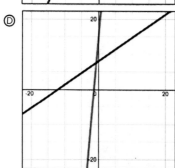

10. **Solve the following system of equations by graphing:**
    2x+3y = 4
    -x+4y = -13

    Ⓐ x = -4, y = 4
    Ⓑ x = 2, y = -2
    Ⓒ x = 5, y = -2
    Ⓓ x = 11, y = -6

Solving Systems of Equations (8.EE.C.8.B)

1. Find the solution to the following system of equations:
   $13x + 3y = 15$ and $y = 5 - 4x$.

   Ⓐ $x = 0, y = 5$
   Ⓑ $x = 5, y = 0$
   Ⓒ $x = 9, y = -31$
   Ⓓ All real numbers are solutions.

2. Solve the system:
   $y = 2x + 5$
   $y = 3x - 7$

   Ⓐ $x = 12, y = 29$
   Ⓑ $x = 3, y = 11$
   Ⓒ $x = 5, y = -2$
   Ⓓ $x = -1, y = 3$

3. Solve the system:
   $2x + 3y = 14$
   $2x - 3y = -10$

   Ⓐ $x = 1, y = 4$
   Ⓑ $x = 2, y = 12$
   Ⓒ $x = 4, y = 2$
   Ⓓ $x = 10, y = 10$

4. Solve the system:
   $x = 13 + 2y$
   $x - 2y = 13$

   Ⓐ $x = 0, y = 13$
   Ⓑ $x = 13, y = 0$
   Ⓒ There is no solution.
   Ⓓ There are infinitely many solutions.

5. Solve the system:
   $2x + 5y = 12$
   $2x + 5y = 9$

   Ⓐ $x = 1, y = 2$
   Ⓑ $x = 2, y = 1$
   Ⓒ There is no solution.
   Ⓓ There are infinitely many solutions.

6. Solve the system:
    -4x + 7y = 26
    4x + 7y = 2

    Ⓐ x = -3, y = 2
    Ⓑ x = 3, y = -2
    Ⓒ x = -2, y = 3
    Ⓓ x = 2, y = -3

7. Find the solution to the following system:
    y + 3x = 11
    y - 2x = 1

    Ⓐ x = -5, y = 2
    Ⓑ x = 2, y = 5
    Ⓒ x = -2, y = -5
    Ⓓ x = -2, y = 5

8. Solve the system:
    2x + 4y = 14
    x + 2y = 7

    Ⓐ x = -1, y = 4
    Ⓑ x = 1, y = 3
    Ⓒ There is no solution.
    Ⓓ There are infinitely many solutions.

9. Solve the system:
    3(y - 2x) = 9
    x - 4 = 0

    Ⓐ x = -4, y = -5
    Ⓑ x = 4, y = 11
    Ⓒ There is no solution.
    Ⓓ There are infinitely many solutions.

10. Solve the system:
    10x = -5(y+2)
    y = 3x-7

    Ⓐ x = -1, y = 4
    Ⓑ x = 2, y = 2
    Ⓒ x = 2, y = -2
    Ⓓ x = 1, y = -4

Systems of Equations in Real-World Problems (8.EE.C.8.C)

1. Jorge and Jillian have cell phones with different service providers. Jorge pays $50 a month and $1 per text message sent. Jillian pays $72 a month and $0.12 per text message sent. How many texts would each of them have to send in order for their bill to be the same amount at the end of the month?

   Ⓐ 2 texts
   Ⓑ 22 texts
   Ⓒ 25 texts
   Ⓓ 47 texts

2. Mr. Stevens is 63 years older than his grandson, Tom. In 3 years, Mr. Stevens will be four times as old as Tom. How old is Tom?

   Ⓐ 17 years
   Ⓑ 18 years
   Ⓒ 20 years
   Ⓓ 22 years

3. Janet has packed a total of 50 textbooks and workbooks in a box, but she can't remember how many of each are in the box. Each textbook weighs 2 pounds, and each workbook weighs 0.5 pounds, and the total weight of the books in the box is 55 pounds. If t is the number of textbooks and w is the number of workbooks, which of the following systems of equations represents this situation?

   Ⓐ t + w = 55
      2t + 0.5w = 50

   Ⓑ 2t + w = 50
      t + 0.5w = 55

   Ⓒ t + w = 50
      2t + 0.5w = 55

   Ⓓ t + w = 55
      2.5(t + w) = 50

4. Plumber A charges $50 to come to your house, plus $40 per hour of labor. Plumber B charges $75 to come to your house, plus $35 per hour of labor. If y is the total dollar amount charged for x hours of labor, which of the following systems of equations correctly represents this situation?

Ⓐ y = 50x + 40
   y = 75x + 35

Ⓑ y = 50x + 40
   y = 35x + 75

Ⓒ y = 40x + 50
   y = 75x + 35

Ⓓ y = 40x + 50
   y = 35x + 75

5. 10 tacos and 6 drinks cost $19.50. 7 tacos and 5 drinks cost $14.25. If t is the cost of one taco and d is the cost of one drink, which of the following systems of equations represents this situation?

Ⓐ 10t + 6d = 19.50
   7t + 5d = 14.25

Ⓑ 6t + 10d = 19.50
   5t + 7d = 14.25

Ⓒ 10t + 7t = 19.50
   6d + 5d = 14.25

Ⓓ 16(t + d) = 19.50
   12(t + d) = 14.25

6. Cindy has $25 saved and earns $12 per week for walking dogs. Mindy has $55 saved and earns $7 per week for watering plants. Cindy and Mindy save all of the money they earn and do not spend any of their savings. After how many weeks will they have the same amount saved? How much money will they have saved?

Ⓐ After 4 weeks, they each will have $83 saved.
Ⓑ After 5 weeks, they each will have $85 saved.
Ⓒ After 6 weeks, they each will have $97 saved.
Ⓓ After 7 weeks, they each will have $104 saved.

7. The seventh and eighth grade classes are raising money for a field trip. The seventh graders are selling calendars for $1.50 each and the eighth graders are selling candy bars for $1.25 each. If they have sold a combined total of 1100 items and each class has the same income, find the number of each item that has been sold.

    Ⓐ 400 calendars and 700 candy bars
    Ⓑ 700 calendars and 400 candy bars
    Ⓒ 500 calendars and 600 candy bars
    Ⓓ 600 calendars and 500 candy bars

8. Anya is three years older than her brother, Cole. In 11 years, Cole will be twice Anya's age now. Find their current ages.

    Ⓐ Anya: 11 years old
       Cole: 8 years old

    Ⓑ Anya: 10 years old
       Cole: 7 years old

    Ⓒ Anya: 9 years old
       Cole: 6 years old

    Ⓓ Anya: 8 years old
       Cole: 5 years old

9. Tom and his sister both decided to get part-time jobs after school at competing clothing stores. Tom makes $15 an hour and recieves $3 in commission for every item he sells. His sister makes $7 an hour and recieves $5 in commission for every item she sells. How many items would each of them have to sell to make the same amount of money in an hour?

    Ⓐ 1 item
    Ⓑ 2 items
    Ⓒ 3 items
    Ⓓ 4 items

10. Lucia and Jack are training for a marathon. Lucia started the first day by running 2 miles and adds 0.25 mile to her distance every day. Jack started the first day by running 0.5 mile and adds 0.5 mile to his distance every day. If both continue this plan, on what day will Lucia and Jack run the same distance?

    Ⓐ Day 3
    Ⓑ Day 6
    Ⓒ Day 9
    Ⓓ Day 12

End of Expressions and Equations

Answer Key and Detailed Explanations
Expressions and Equations
Properties of Exponents (8.EE.A.1)

Question No.	Answer	Detailed Explanation
1	D	Unless there are parentheses to denote otherwise, the exponent is only applied to the constant or variable immediately preceding it. If there are parentheses immediately preceding it, then it is applied to everything within the parentheses.
2	D	When dividing quantities with like bases, you must subtract the exponents. $6 - (-2) = 8$ $$\frac{X^6}{X^{-2}} = X^{6-(-2)} = X^8$$
3	A	A quick way to change exponents from negative to positive is to move the expression to which the negative exponent is applied from the denominator to the numerator or vice versa and change the sign of the exponent. 3^{-2} is the same as $\frac{1}{3^2}$ which is the same as $\frac{1}{9}$
4	C	$X^{(2-5)} = X^{-3}$ Now move x-3 to the denominator and change the sign of the exponent from negative to positive. $\frac{1}{X^3}$
5	A	Regardless of the number of 1s that we multiply the result is always 1 because 1 is the identity element for multiplication.
6	C	When multiplying quantities with the same base, you add exponents. $(X^{-3})(X^{-3}) = X^{-6}$ To change the exponent -6 to positive 6, you write the reciprocal of X^{-6}. $\frac{1}{X^6}$
7	B	$(x^{-2})^{-7} = x^{14}$ because to raise a power to a power, we multiply exponents.
8	C	$4 \times 0 = 0$ and any quantity to the 0 power is 1 by definition.
9	B	When raising a power to a power, multiply exponents.

Question No.	Answer	Detailed Explanation
10	A	$5^2 = 25$ and $5^3 = 125$ $25 + 125 = 150$

Square & Cube Roots (8.EE.A.2)

1	A	$10^3 = 1,000$, So, the cube root of 1,000 is 10.
2	C	$8\sqrt{12} = 8\sqrt{(4 \times 3)} = 8(2)\sqrt{3} = 16\sqrt{3}$ $\sqrt{15} = \sqrt{(5 \times 3)}$ So $8\sqrt{12} \div \sqrt{15} = 16\sqrt{3} \div \sqrt{5}\sqrt{3}$
3	A	$8^2 = 64$ and $9^2 = 81$ Therefore, the square root of 75 is between 8 and 9.
4	A	$10^2 = 100$ and $11^2 = 121$ Therefore, the square root of 110 is between 10 and 11.
5	A	$6\sqrt{20} \div \sqrt{5} = 6\sqrt{4}\sqrt{5} \div \sqrt{5} = 6\sqrt{4} = 12$
6	A	$4^3 = 64$ and $5^3 = 125$ Therefore, the cube root of 66 is between 4 and 5.
7	A	$3\sqrt{144} \div \sqrt{12} = 3(12) \div \sqrt{12} = 36 \div \sqrt{12}$ Note that this expression may be further reduced.
8	C	$7^3 = 343$ and $8^3 = 512$ Therefore, the cube root of 400 lies between 7 and 8.
9	D	$4\sqrt{250} \div 5\sqrt{2} = 4\sqrt{25}\sqrt{10} \div 5\sqrt{2} = 4(5)\sqrt{10} \div 5\sqrt{2} = 4\sqrt{10} \div \sqrt{2} = 4\sqrt{5}\sqrt{2} \div \sqrt{2} = 4\sqrt{5}$
10	C	$15^3 = 3,375$; $10^3 = 1,000$; $5^3 = 125$; $3^3 = 27$ 150 is closest to 5^3 or 125.

Scientific Notation (8.EE.A.3)

1	B	Moving the decimal in 3,380,000 just to the right of the first non-zero digit requires us to move it 6 places to the left resulting in 3.38×10^6.
2	C	If I move the decimal from standard notation to the right of the first non-zero digit, I must move it 8 places to the left. Therefore, 890,800,000 expressed in scientific notation is, 8.908×10^8.
3	B	Moving the decimal to the right 6 places from 3.8×10^6, we get 3,800,000.
4	C	From 1.53×10^7, we move the decimal to the right 7 places giving us 15,300,000.

Question No.	Answer	Detailed Explanation
5	B	Move the decimal 3 places to the left.
6	C	$1 \leq N \leq 9.99...$ and we must move the decimal 5 places to the right resulting in $100,000 \leq N < 1,000,000$.
7	C	To change to standard notation, I must move the decimal 9 places to the left because the exponent is negative. The result is 0.000000650. Changing to scientific notation we get 6.5×10^{-7}. Or, since we must move the decimal 2 places to the left to get it properly placed for scientific notation, we only need to move it 7 more places for standard notation; so scientific notation would be written as 6.50×10^{-7}.
8	C	From 1.0, I must move the decimal 4 places to the left resulting in 0.0001. Since the 1 is in the ten-thousandths place, the fraction will be $\frac{1}{10,000}$
9	C	$8.935 \times 10^3 = 8,935$ which is the greatest of the numbers shown.
10	B	Since we have negative numbers and positive numbers, the lowest value will be negative. -1,560 is less than -156.

Solving Problems Involving Scientific Notation (8.EE.A.4)

1	B	3.7×10^7 people $\div 1.6 \times 10^5$ square miles $=(3.7 \div 1.6) \times 10^{7-5} = 2.3125 \times 10^2$ people/square mile = 231.25 people/sq. mi. 240 is the best estimate.
2	C	6×10^7 kilometers $\div 3 \times 10^5$ kilometers per second $= (6 \div 3) \times 10^{(7-5)} = 2 \times 10^2$(kilometers $\div$ kilometers/seconds) $= 2 \times 10^2$ seconds = 200 seconds
3	A	$(4 \times 10^6) \times (2 \times 10^3) = 8 \times 10^{(6+3)} = 8 \times 10^9$
4	D	$(2 \times 10^{-3}) \times (3 \times 10^5) = (2 \times 3) \times 10^{(-3+5)} = 6 \times 10^2$
5	B	$(2.4 \times 10^3) / (6 \times 10) = 0.4 \times 10^2 = 4 \times 10 = 40$
6	A	$(5 \times 105) \times (9 \times 10-3) = (45 \times 105-3) = 45 \times 102 = 4.5 \times 103$ Therefore, 4.5 x 104 is not equal to 4.5 x 103, so 4.5 x 104 is the correct answer.

Question No.	Answer	Detailed Explanation
7	A	$(5 \times 10^7) \div (10 \times 10^2) = 0.5 \times 10^{(7-2)} = 0.5 \times 10^5 = 5 \times 10^4$
8	C	$.00004567 \times .00001234 = (4.567 \times 10^{-5}) \times (1.234 \times 10^{-5}) = (4.567 \times 1.234) \times 10^{(-5 + -5)} = 5.635678 \times 10^{-10}$ 5.636×10^{-10} is the correct answer.
9	D	$(50.67 \times 10^4) \times (12.9 \times 10^3) = 653.643 \times 10^{(4+3)} = 653.643 \times 107 = 6,536,430,000$ $6,536,000,000$ is the correct answer.
10	A	$(1.298 \times 10^4) \div (3.97 \times 10^2) = 0.326952 \times 10^2 = 32.7$ 32.7 is the best answer.

Compare Proportions (8.EE.B.5)

1	C	Unit rate means cost per unit (one). "Per" means to divide; so we divide the total cost by the number of units. $\$1440 \div 12 = \$120$ per unit
2	B	$\$2.90 \div 10 = \$0.29$ per piece $\$6.20 \div 20 = \$0.31$ per piece Therefore, Fruity is $\$0.02$ more per piece than Big Bubbles.
3	A	Slope is defined as change in vertical height per unit change in horizontal distance. 9 ft $\div$ 54 ft = 1/6; 12 ft $\div$ 84 ft = 1/7 Since 1/6 > 1/7, the first slope is steeper.
4	B	4 ft $\div$ 12 ft = 1/3 which is the largest ratio so the steepest slope.
5	A	The unit rate can also be called the average or mean, but not the mode, median, nor the frequency.
6	A	Unit cost is calculated by dividing the total cost by the amount of items. This statement is true. Therefore, "Unit cost is calculated by dividing the amount of items by the total cost." must be a false statement. (This would represent the number of items you get per dollar paid.)

Question No.	Answer	Detailed Explanation
7	D	1 qt = 32 fl oz 1 gallon = 128 fl oz 5 gallons = 640 fl oz $3.99/24 fl oz = $0.16625 per fl oz $4.80/32 fl oz = $0.15 per fl oz $11.00/128 fl oz = $0.0859375 per fl oz $49.60/640 fl oz = $0.0775 per fl oz--- So, the five gallon tub offers the best price
8	D	5 ft/30 ft = 1/6 which is the least steep slope
9	B	6 tee shirt for $90 $90 ÷ 6 = $15 per tee shirt, which is the most expensive of the four options.
10	A	15 minutes ÷ 5 laps = 3 min/lap which is the fastest time
11	B	$2.00 for a 6 pack-----$0.30 per can $6.00 for a 24 pack-----$0.25 per can is the lowest unit cost $3.60 for a 12 pack------$0.30 per can $10.00 for a 36 pack------$0.36 per can
12	B	Plain wafers have 20 per pack and sell for $2.40------That is $0.12 per wafer. Sugar-free wafers have 30 pieces per pack and sell for $6.30-----That is $0.21 per wafer. ---------$0.09 more per unit than the plain wafers
13	A	36 ÷ 40 = 90% which is the highest score of the four practice tests.
14	D	$30 ÷ 40 tickets = $0.75 per ticket
15	C	Ski Slope A has a vertical rise of 4 feet and a horizontal run of 16 feet---1/4 Ski Slope B has a vertical rise of 3 feet and a horizontal run of 12 feet---1/4 Ski Slope C has a vertical rise of 3 feet and a horizontal run of 9 feet---1/3 has the steepest slope. Ski Slope D has a vertical rise of 5 feet and a horizontal run of 25 feet----1/5

Understanding Slope (8.EE.B.6)

Question No.	Answer	Detailed Explanation
1	D	Only the last statement, "Slope is determined by dividing the vertical distance between two points by the corresponding horizontal distance," is true.
2	D	The slope of the line passing through these two points is (4 - 2) / (- 1 - 1) = 2/- 2 = - 1. The only equation with a slope of - 1 is y = - x + 3.
3	D	The slope of y = 4x + 3 is 4. The slope of y = 4x - 2 is 4.
4	D	$y = \frac{9}{5}x - 3$ has a slope of 9/5 which is the greatest.
5	A	1/8 is the smallest slope. $y = \frac{1}{8}x + 7$ is the correct answer.
6	A	If we know the y-intercept and the slope, we can write the equation of a straight line. If we know the y-intercept, we know b in the slope-intercept formula. The y-intercept together with another point or the x-intercept make it possible to determine the slope of the line. Direction would not give enough information
7	C	Slope = rise/run (vertical change/horizontal change) Slope = 30/500 = 3/50
8	C	y = mx + b m = slope In this case, we have y = (3/2)x -9 so m = 3/2
9	A	We know the y-intercept is (0, 5) and can determine the slope is 10/2=5, giving us the equation of the line: y=5x+5. We can check by substituting the coordinates for x and solve for y. (0,5): y = 5x + 5 = 5(0) + 5 = 5 = y (2,15): y = 5x + 5 = 5(2) + 5 = 10 + 5 = 15 = y
10	A	m = (10-6)/(3-1)=4/2=2 The equation where the slope equals 2 is y = 2x - 9.

Question No.	Answer	Detailed Explanation
11	A	The slope, m=(10-6)/(5-3) is 2. The equation where the slope equals 2 is y = 2x - 12
12	B	The equation y = -5x - 4 has -5 as its slope, and The slope of y = -5x - 11 is also -5.
13	B	(5 - 3) ÷ (5 - 3) = 1 The slope is 1.
14	A	The slopes of these lines are 4, -3, and 3 respectively. Of these, 4 would be the steepest slope.
15	C	(0,-3) and (-3,0) The slope of the line passing through these two points is (0 - - 3) ÷ (- 3 - 0) = 3 ÷ - 3 = - 1. We know that the y-intercept is at (0, - 3) so the line is y = -x - 3.

Solving Linear Equations (8.EE.C.7.A)

1	A	Consecutive odd integers lie 2 units apart on the number line; so let the numbers be represented by n and n+2. Their sum is 44; so n + n + 2 = 44. 2n +2 = 44 2n = 44 - 2 2n = 42 n = 21; so n + 2 = 23
2	A	Cumulative GPA =(2.1 +2.9 +3.1 + n)/4 3.0 =(8.1 + n)/4 12.0 = 8.1 + n 3.9 = n
3	C	Let m = miles/day for remaining 3 days m = (original miles - miles already traveled) / 3 m = (1924 - 490) / 3 = 1434 / 3 = 478 miles per day
4	C	3x + 5 = 29 3x = 29 - 5 3x = 24 x = 8
5	D	7 - 2x = 13 -2x 7 - 13 = 0 - 6 = 0 Since this is a false statement, there is not solution to this equation.

Question No.	Answer	Detailed Explanation
6	B	$6x + 1 = 4x - 3$ $6x - 4x = -3 - 1$ $2x = -4$ $x = -4/2$ $x = -2$
7	D	$2(x + 3) + 1 = 7 + 2x$ $2x + 6 + 1 = 7 + 2x$ $2x + 7 = 7 + 2x$ Since both sides of the equation are identical, any real solution satisfies the equation.
8	D	$8x - 1 = 8x$ $-1 = 0$ $7x - 8 = 8x$ $-8 = 8x - 7x$ $-8 = x$ Since this is a false statement, there is not solution to this equation.
9	D	$2x + 5x - 9 = 8x - x - 3 - 6$ $7x - 9 = 7x - 9$ All real values for x are solutions.
10	D	$-y + 7y - 54 = 0$ $6y - 54 = 0$ $6y = 54$ $y = 9$

Solve Linear Equations with Rational Numbers (8.EE.C.7.B)

1	C	$7/14 = n + (7/14)(n)$ $1/2 = n + (1/2)n$; Multiply by 2 to eliminate denominators. $1 = 2n + n$ $1 = 3n$ $1/3 = n$
2	B	$2(2x - 7) = 14$ $4x - 14 = 14$ $4x = 14 + 14$ $4x = 28$ $x = 7$
3	D	$6x - (2x + 5) = 11$ $6x - 2x - 5 = 11$ $4x - 5 = 11$ $4x = 11 + 5$ $4x = 16$ $x = 4$

Question No.	Answer	Detailed Explanation
4	B	$4x + 2(x - 3) = 0$ $4x + 2x - 6 = 0$ $6x - 6 = 0$ $6x = 6$ $x = 6 \div 6$ $x = 1$
5	A	$3y - 7(y + 5) = y - 35$ $3y - 7y - 35 = y - 35$ $- 4y - 35 = y - 35$ $- 35 = y - 35 + 4y$ $- 35 + 35 = y + 4y$ $0 = 5y$ $0 = y$
6	A	$2(x-5) = 1/2 (6x + 4)$ $2x - 10 = 3x + 2$ $-12 = x$
7	D	$3x + 2 + x = 1/3 (12x + 6$ $3x + x = 4x + 2$ $4x + = 4x + 2$ $0 = 0$ Thus, all real values of x are correct solutions.
8	C	$\frac{1}{2} x + \frac{2}{3} x + 5 = \frac{5}{2} x + 6$ $6(\frac{1}{2}x + \frac{2}{3} x + 5 = \frac{5}{2} x + 6)$ $3x + 4x + 30 = 15x + 36$ $-8x = 6$ $x = - \frac{6}{8} = - \frac{3}{4}$
9	B	$2(2x + 3) = 3(2x + 5)$ $4x + 6 = 6x + 15$ $-2x + 6 = 15$ $-2x = 9$ $x = - \frac{9}{2}$
10	B	$.64x - .15x + .08 = .09x$ $.49x + .08 = .09x$ $.08 = - .04x$ $- .2 = x$

Solutions to Systems of Equations (8.EE.C.8.A)

Question No.	Answer	Detailed Explanation
1	D	x - 5 = 2x + 1 - 5 - 1 = 2x - x -6 = x Then y = x - 5 y = - 6 - 5 y = - 11
2	C	These lines have the same slope, but different y-intercepts, so they do not intersect. There is no solution. The lines are parallel.
3	D	y = 2(2 - 3x) = - 6x + 4 y = -3(2x + 3) = - 6x - 9 There is no solution because for any value for x, we have 2 different values for y. This is inconsistent.
4	B	$\frac{x}{2} + \frac{y}{2} = 2$ 3x - 2y = 48 Multiplying the first equation by 6 gives 3x + 2y = 12 Adding to 3x - 2y = 48 gives 6x = 60 x = 10 3(10) - 2y = 48 30 - 2y = 48 -2y = 48 - 30 = 18 y = -9
5	B	y = 2x - 6 y = -2x + 6 2x - 6 = - 2x + 6 2x + 2x = 6 + 6 4x = 12 x = 3 y = 2(3) -6 = 0
6	C	y = -x + 3:m = -1, b = 3 y = -2x + 8: m = -2b, b = 8 Solution (5,2)
7	A	2x + y = 12 : m = -2, b = 12 $y = \frac{1}{5} x : m = \frac{1}{5}$, b = 0 Solution$(\frac{12}{11}, \frac{60}{11})$

Question No.	Answer	Detailed Explanation
8	C	A system of equaitons has infinite solutions when both lines are the same.
9	A	A system of equations has no solutions when both lines are parallel because they will never intersect.
10	C	$2x + 3y = 4 : x - \text{int} = 2, y - \text{int} = \frac{4}{3}$ $-x + 4y = -13 : x -\text{int} = 13, y\text{-int} = \frac{13}{4}$ Solution (5, -2)

Solving Systems of Equations (8.EE.C.8.B)

1	A	$13x + 3y = 15$ $\underline{4x + y = 5}$ Multiply bottom equation by - 3 and $\underline{-12x - 3y = -15}$; Add first and last equations. $x = 0$ $x = 0$ Substitute into $y = 5 - 4x$ and $y = 5 - 0$ $y = 5$
2	A	$y = 2x + 5$ $y = 3x - 7$ $2x + 5 = 3x - 7$ $5 = 3x - 7 - 2x$ $5 + 7 = 3x - 2x$ $12 = x$ $y = 2(12) + 5 = 24 + 5 = 29$
3	A	Adding the two equations, we get $4x = 4$; $x =1$ Then substituting into $2x + 3y = 14$, we get $2 + 3y = 14$ $3y = 14 - 2$ $3y = 12$ $y = 4$
4	D	Solving the second equation for x, we get $x = 13 + 2y$ which is identical to the first equation. Therefore, there are infinitely many points (x, y) that satisfy the system.

Question No.	Answer	Detailed Explanation
5	C	Examining these equations we find an inconsistency. $2x + 5y$ CANNOT be equal to two different quantities. Therefore, there is no solution.
6	A	Adding the two equations, we get $14y = 28$. $y = 2$ Substituting into $4x + 7y = 2$, we get $4x + 14 = 2$ $4x = -12$ $x = -3$
7	B	$y + 3x = 11$ $y - 2x = 1$ $2y + 6x = 22$ $3y - 6x = 3$ $5y = 25$ $y = 5$ $x = 2$
8	D	$2x + 4y = 14$ $x + 2y = 7$ Dividing the first equation by 2, we get the second equation. Therefore, there are an infinite number of solutions.
9	B	$3(y - 2x) = 9$ $x - 4 = 0$ $x = 4$ $3y - 6x = 9$ $3y - 6(4) = 9$ $3y - 24 = 9$ $3y = 9 + 24$ $3y = 33$ $y = 11$
10	D	$10x = -5((3x - 7)+2)$ $10x = -5(3x - 5)$ $10x = -15 + 25$ $25x = 25$ $x = 1$ $y = 3(1) - 7$ $y = -4$

Systems of Equations in Real-World Problems (8.EE.C.8.C)

Question No.	Answer	Detailed Explanation
1	C	Jorge: y = 50 + 1x Jillian: y = 72 + .12x 50 + 1x = 72 + .12x .88x = 22 x = 25
2	B	Now: Tom = x; Mr. Stevens = y and y = x + 63 In 3 Yrs: Tom = x + 3; Mr. Stevens = y + 3 and y + 3 = 4(x + 3) Simplify y + 3 = 4(x + 3) y = 4x + 12 - 3 y = 4x + 9 and then our system of equations is y = x + 63 and y = 4x + 9 Multiply top equation by - 4 and - 4y = -4x - 252; Add y = 4x + 9 $\overline{-3y = - 243}$ y = 81 Substitute into y = x + 63 81 = x + 63 and 18 = x which is Tom's age The correct answer is 18 years.
3	C	There are a total of 50 textbooks and workbooks in the box; so t + w = 50. Each textbook weighs 2 lb and each workbook weighs 0.5 lb and the total weight is 55 lb; so 2t + 0.5w = 55.
4	D	Total cost for Plumber A is 40x + 50 and for Plumber B is 35x + 75.
5	A	10t + 6d = 19.50 7t + 5d = 14.25 correctly expresses the relative costs as described.

Question No.	Answer	Detailed Explanation
6	C	Let C = money Cindy has. Let M = money Mindy has. C = 25 + 12w where w is the weeks worked. M = 55 + 7w where w is the weeks worked. 25 + 12w = 55 + 7w 12w - 7w = 55 - 25 5w = 30 w = 6 weeks C = 25 + 12(6) = \$97.00 M = 55 + 7(6) = \$97.00
7	C	Let x = calendars and y = candy bars x + y = 1100 1.50x = 1.25y 1.50x + 1.50y = 1.50(1100) 1.50x + 1.50y = 1650 1.50x - 1.25y = 0; Subtracting we get 2.75y = 1650 y = 600 candy bars sold x = 1100 - y = 500 calendars
8	D	Let x = Cole's age now and y = Anya's age now. y = x + 3 x + 11 = 2y 2y - x = 11 y - x = 3 -y + x = -3 y = 8 years old x = 5 years old
9	D	Tom: y = 15 + 3x Sister: y = 7 + 5x 15 + 3x = 7 + 5x 8 = 2x x = 4
10	B	Lucia: y = 2 + .25x Jach: y = .5 + .5x 2 + .25x = .5 + .5x 1.5 = .25x 6 = x

Functions

Functions (8.F.A.1)

1. Which of the following is NOT a function?

    Ⓐ {(2, 3), (4, 7), (8, 6)}
    Ⓑ {(2, 2), (4, 4), (8, 8)}
    Ⓒ {(2, 3), (4, 3), (8, 3)}
    Ⓓ {(2, 3), (2, 7), (8, 6)}

2. Which of the following tables shows that y is a function of x?

Ⓐ
x	y
1	4
1	7
4	7

Ⓑ
x	y
1	7
3	8
4	7

Ⓒ
x	y
3	2
3	7
4	8

Ⓓ
x	y
1	7
4	7
4	9

3. If y is a function of x, which of the following CANNOT be true?

    Ⓐ A particular x value is associated with two different y values.
    Ⓑ Two different x values are associated with the same y value.
    Ⓒ Every x value is associated with the same y value.
    Ⓓ Every x value is associated with a different y value.

4. Each of the following graphs consists of two points. Which graph could NOT represent a function?

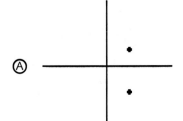

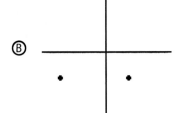

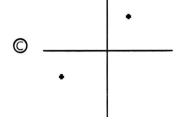

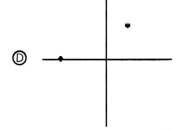

5. Which of the following could NOT be the graph of a function?

Ⓐ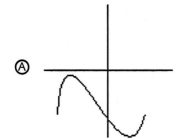

Ⓑ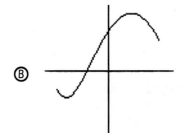

Ⓒ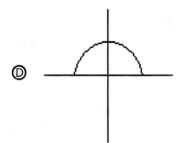

Ⓓ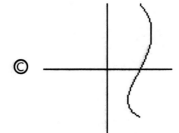

6. Which of the following is NOT a function?

Ⓐ {(0, 0), (2, 2), (4, 4)}
Ⓑ {(0, 4), (2, 4), (4, 4)}
Ⓒ {(0, 0), (2, 0), (4, 0)}
Ⓓ {(0, 0), (0, 2), (0, 4)}

7. Which of the following is true of the graph of any non-constant function?

    Ⓐ A line drawn parallel to the x-axis will never cross the graph.
    Ⓑ A line drawn perpendicular to the x-axis will never cross the graph.
    Ⓒ A line drawn through the graph parallel to the x-axis will cross the graph one and only one time.
    Ⓓ A line drawn through the graph perpendicular to the x-axis will cross the graph one and only one time.

8. The given set represents a function:
    {(0,1), (1,1), (2,1)}
    If the ordered pair ____ was added to the set, it would no longer be a function.

    Ⓐ (3,1)
    Ⓑ (3,2)
    Ⓒ (3,3)
    Ⓓ (2,3)

9. In order for this set, {(6,5), (5, 4), (4,3)}, to remain a function, which of the following ordered pairs COULD be added to it?

    Ⓐ (6,6)
    Ⓑ (5,5)
    Ⓒ (4,4)
    Ⓓ (3,3)

10. Which of the following sets of ordered pairs represents a function?

    Ⓐ [(0,1), (0,2), (1,3), (1,4)]
    Ⓑ [(1,1), (1,2), (1,3), (1,4)]
    Ⓒ [(2,5), (2,6), (4,7), (5,7)]
    Ⓓ [(-7,10), (7,10), (8,9), (9,10)]

Comparing Functions (8.F.A.2)

1. A set of instructions says to subtract 5 from a number n and then double that result, calling the final result p. Which function rule represents this set of instructions?

    Ⓐ p = 2(n – 5)
    Ⓑ p = 2n – 5
    Ⓒ n = 2(p – 5)
    Ⓓ n = 2p – 5

2. Which of the following linear functions is represented by the (x, y) pairs shown in the table below?

x	y
-3	-1
1	7
4	13

    Ⓐ y = x + 2
    Ⓑ y = 2x + 5
    Ⓒ y = 3x + 1
    Ⓓ y = 4x + 3

3.

x	y
0	3
1	5
2	7

    Three (x, y) pairs of a linear function are shown in the table above. Which of the following functions has the same slope as the function shown in the table?

    Ⓐ y = 3x + 2
    Ⓑ y = 2x -4
    Ⓒ y = x + 3
    Ⓓ y = x - 1

4. The graph of linear function A passes through the point (5, 6). The graph of linear function B passes through the point (6, 7). The two graphs intersect at the point (2, 5). Which of the following statements is true?

   Ⓐ Function A has the greater slope.
   Ⓑ Both functions have the same slope.
   Ⓒ Function B has the greater slope.
   Ⓓ No relationship between the slopes of the lines can be determined from this inforation.

5. If line M includes the points (-1, 4) and (7, 9) and line N includes the points (5, 2) and (-3, -3), which of the following best describes the relationship between M and N?

   Ⓐ They are perpendicular
   Ⓑ They intersect but are not perpendicular
   Ⓒ They are parallel
   Ⓓ Not enough information is provided

6. If line R includes the points (-2, -2) and (6,4) and line S includes the points(0,4) and (3,0) which of the following best describes the relationship between R and S?

   Ⓐ They are perpendicular.
   Ⓑ They intersect but are not perpendicular.
   Ⓒ They are parallel.
   Ⓓ Not enough information is provided.

7.

x	y
-4	13
2	1
6	-7

Three points of a linear function are shown in the table above. What is the y-intercept of this function?

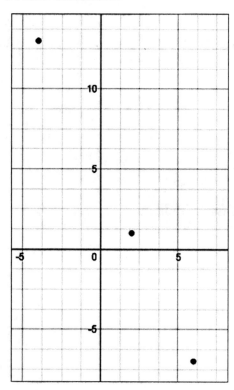

Ⓐ (0, 5)
Ⓑ (0, 6)
Ⓒ (0, 7)
Ⓓ (0, 8)

8. Comparing the two linear functions, y = 2x + 7 and y = 7x + 2, we find that

Ⓐ y = 2x + 7 has the steeper slope.
Ⓑ The graphs of the functions will be parallel lines.
Ⓒ The graphs of the functions will intersect at the point (1, 9).
Ⓓ The graphs of the functions will be perpendicular lines.

9. Which of the following is true of the graphs of the lines y = 3x +5 and y = -x/3 +6?

    Ⓐ The graphs will be two parallel straight lines.
    Ⓑ The graphs will be two perpendicular straight lines.
    Ⓒ Both functions will produce the same line.
    Ⓓ There is not enough information to compare.

10. If line A, denoted by y = x + 9, and line B, denoted by y = 5x - 3 are graphed, which of the following statements is correct?

    Ⓐ Line B has the steeper slope.
    Ⓑ Both lines have the same slope.
    Ⓒ Line A has the steeper slope.
    Ⓓ There is not enough information to answer the question.

Linear Functions (8.F.A.3)

1. A linear function includes the ordered pairs (2, 5), (6, 7), and (k, 11). What is the value of k?

   Ⓐ 8
   Ⓑ 10
   Ⓒ 12
   Ⓓ 14

2. Which of the following functions is NOT linear?

   Ⓐ $f(x) = x + 0.5$
   Ⓑ $f(x) = -x + 0.5$
   Ⓒ $f(x) = x^2 + 0.5$
   Ⓓ $f(x) = 0.5x$

3. Which of the following functions is linear and includes the point (3, 0)?

   Ⓐ $f(x) = 3/x$
   Ⓑ $f(x) = x - 3$
   Ⓒ $f(x) = 3$
   Ⓓ $f(x) = 3x$

4. Four (x, y) pairs of a certain function are shown in the table below. Which of the following best describes the function?

x	y
-3	1
-1	4
1	7
3	10

   Ⓐ The function increases linearly.
   Ⓑ The function decreases linearly.
   Ⓒ The function is constant.
   Ⓓ The function is not linear.

5. Four (x, y) pairs of a certain function are shown in the table below. Which of the following statements most accurately describes the function?

x	y
0	3
1	4
2	7
3	12

Ⓐ The function is linear because it does not include the point (0, 0).
Ⓑ The function is linear because it does not have the same slope between different pairs of points.
Ⓒ The function is nonlinear because it does not include the point (0, 0).
Ⓓ The function is nonlinear because it does not have the same slope between differnt pairs of points.

6. The graph of a linear function lies in the first and fourth quadrants. Which of the following CANNOT be true?

Ⓐ It is an increasing function.
Ⓑ It is a constant function.
Ⓒ It also lies in the second quadrant.
Ⓓ It also lies in the third quadrant.

7. A linear function includes the ordered pairs (0, 1), (3, 3), and (9, n). What is the value of n?

Ⓐ 5
Ⓑ 6
Ⓒ 7
Ⓓ 8

8. A linear function includes the ordered pairs (0, 3), (3, 9), and (9, n). What is the value of n?

Ⓐ 20
Ⓑ 14
Ⓒ 21
Ⓓ 22

9. The graph of a linear function with a non-negative slope lies in the first and second quadrants. Which of the following CANNOT be true?

    Ⓐ It is an increasing function.
    Ⓑ It is a constant function.
    Ⓒ It also lies in the third quadrant.
    Ⓓ It also lies in the fourth quadrant.

10. Which function is represented by this table?

x	y
0	2
1	4
2	6
3	8

    Ⓐ y = 2x + 2 ✓
    Ⓑ y = 3x - 4
    Ⓒ y = 4x - 5
    Ⓓ y = 6x - 8

Linear Function Models (8.F.A.4)

1. If a graph includes the points (2, 5) and (8, 5), which of the following must be true?

   Ⓐ It is the graph of a linear function.
   Ⓑ It is the graph of an increasing function.
   Ⓒ It is not the graph of a function.
   Ⓓ None of the above

2. The graph of a linear function y = mx + 2 goes through the point (4, 0). Which of the following must be true?

   Ⓐ m is negative.
   Ⓑ m = 0
   Ⓒ m is positive
   Ⓓ Cannot be determined.

3. The graph of a linear function y = 2x + b passes through the point (-5, 0). Which of the following must be true?

   Ⓐ b is positive.
   Ⓑ b is negative.
   Ⓒ b = 0
   Ⓓ Cannot be determined.

4. The graph of a linear function y = mx + b goes through the point (0, 0). Which of the following must be true?

   Ⓐ m is positive.
   Ⓑ m is negative.
   Ⓒ m = 0
   Ⓓ b = 0

5. The graph of a linear function y = mx + b includes the points (2, 5) and (9, 5). Which of the following gives the correct values of m and b?

   Ⓐ m = 5 and b = 7
   Ⓑ m = 1 and b = 0
   Ⓒ m = 0 and b = 5
   Ⓓ m = 2 and b = 9

6. **What is the slope of the linear function represented by the (x, y) pairs shown in the table below?**

x	y
0	11
2	7
3	5

Ⓐ $-\dfrac{1}{2}$

Ⓑ $\dfrac{1}{2}$

Ⓒ -2

Ⓓ 2

7. **The graph of a certain linear function includes the points (-4, 1) and (5, 1). Which of the following statements most accurately describes the function?**

Ⓐ It is an increasing linear function.
Ⓑ It is a decreasing linear function.
Ⓒ It is a constant function.
Ⓓ It is a nonlinear function.

8. **Which of the following linear functions has the greatest slope?**

Ⓐ
x	y
0	1
2	5
4	9

Ⓑ
x	y
0	3
2	6
4	9

Ⓒ
x	y
0	5
2	7
4	9

Ⓓ
x	y
0	7
2	8
4	9

9. A young child is building a tower of blocks on top of a bench. The bench is 18 inches high, and each block is 3 inches high. Which of the following functions correctly relates the total height of the tower (including the bench) h, in inches, to the number of blocks b?

Ⓐ h = 3b - 18
Ⓑ h = 3b + 18
Ⓒ h = 18b - 3
Ⓓ h = 18b + 3

10. Jim owes his parents $10. Each week, his parents pay him $5 for doing chores. Assuming that Jim does not earn money from any other source and does not spend any of his money. Which of the following functions correctly relates the total amount of money m, in dollars, that Jim will have to the number of weeks w?

Ⓐ m = -5w - 10
Ⓑ m = -5w + 10
Ⓒ m = 5w - 10
Ⓓ m = 5w + 10

11. An amusement park charges $5 admission and an additional $2 per ride. Which of the following functions correctly relates the total amount paid p, in dollars, to the number of rides r?

Ⓐ p = 2r + 5
Ⓑ p = 5r + 2
Ⓒ p = 10r
Ⓓ p = 7r

12. Which of the following linear functions has the smallest slope?

Ⓐ

x	y
1	2
4	6
7	8

Ⓒ

x	y
1	4
4	6
7	8

Ⓑ

x	y
1	3
4	6
7	9

Ⓓ

x	y
1	5
4	6
7	7

13. The graph of a linear function includes the points (6, 2) and (9, 4). What is the y in tercept of the graph?

Ⓐ (0, 0)
Ⓑ (0, -2)
Ⓒ (3, 0)
Ⓓ (0, 3)

14. Which of the following functions has this set of points as solutions?
   {(0, -5), (1, 0), and (4, 15)}

Ⓐ f(x) = 4x - 15
Ⓑ f(x) = 0
Ⓒ f(x) = -5
Ⓓ f(x) = 5x - 5

15. A music store is offering a special on CDs. The cost is $20.00 for the first CD and $10.00 for each additional CD purchased.
   Which of the following functions represents the total amount in dollars of your purchase where x is the number of CDs purchased?

Ⓐ f(x) = 10x + 20
Ⓑ f(x) = 10(x - 1) + 20
Ⓒ f(x) = 20x + 10
Ⓓ f(x) = 20(x - 1) + 10

Analyzing Functions (8.F.B.5)

1. Complete the following:
   The cost per textbook is a function of the number of copies of any one title purchased. This implies that _____

   Ⓐ the cost per copy of any one title is always a constant.
   Ⓑ the cost per copy of any one title will change based on the number of copies pur chased.
   Ⓒ the cost per copy of any one title is not related to the number of copies purchased.
   Ⓓ None of the above.

2. If a student's math grade is a positive function of the number of hours he spends pre paring for a test, which of the following is correct?

   Ⓐ The more he studies, the lower his grade.
   Ⓑ The more he studies, the higher his grade.
   Ⓒ There is no relation between how much he studies and his grade.
   Ⓓ The faster he finishes his work, the higher his grade will be.

3. Mandy took a math quiz and received an initial score of i. She retook the quiz several times and, with each attempt, doubled her previous score.
   After a TOTAL of four attempts, her final score was _____.

   Ⓐ 2i
   Ⓑ 3i
   Ⓒ 2^3i
   Ⓓ None of the above.

4. Which story matches the graph below?

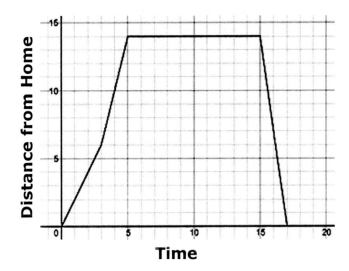

Ⓐ Kevin took his dog for a walk. He started slowly, but then increase his speed. Kevin sat on a bench while his dog played around in the park. Then, he realized it was getting late and walked home quickly.

Ⓑ Kevin rode his bicycle up a really steep hill. At the top of the hill, the road flattened out. Then, he flew down the other side.

Ⓒ Kevin went shopping with his mom. She slowly pulled out of the driveway and quickly approached the store. Kevin and his mom parked the car, quickly did their shopping, and went home.

Ⓓ Kevin went for a run. He started off at good pace and gained some speed until he ran out of breath. After catching his breath, Kevin decided to just walk home.

5. Which story matches the graph below?

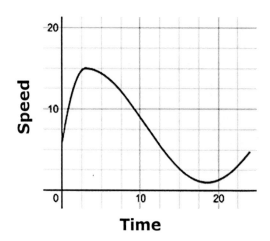

Time

Ⓐ Alison went sledding with her friends. In her excitement, she ran up the hill. When she reached the top, she jumped on her sled and quickly raced to the bottom.

Ⓑ Alison started at a stoplight and rapidly increased her speed. She noticed the stoplight up ahead was red, so she braked to slow down. As she approached the light, it turned green so she increased her speed again.

Ⓒ Alison decided to go bungee jumping. She jumped of the platform, speeding towards the ground. Then she stopped for just a moment before the bungee pulled her back up.

Ⓓ Alison is running for student government and has decided to make 200 buttons. At first, she is really excited and makes buttons at a very quick rate. However, she gets bored with the process and her hand starts to cramp. She eventually gives up on the project.

6. Use the graph below to answer the following question. If each runner continues at their current pace, which runner will win the race?

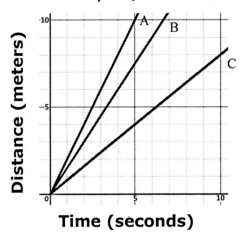

Time (seconds)

Ⓐ Runner A
Ⓑ Runner B
Ⓒ Runner C
Ⓓ Runner A and B

7. Given the graph below, what is the speed of Runner A?

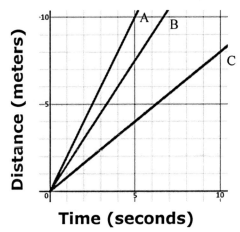

Time (seconds)

Ⓐ 2 m/s
Ⓑ 1.5 m/s
Ⓒ 0.8 m/s
Ⓓ 0.6 m/s

8. The following graph shows the change in water level as Karina takes a bath. What portion of the graph corresponds to when Karina is relaxing in the tub?

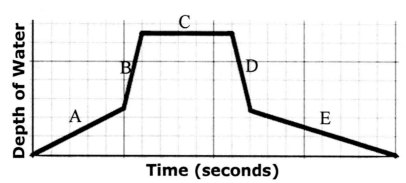

Ⓐ A
Ⓑ B
Ⓒ C
Ⓓ D

9. What function includes the line segment graphed below?

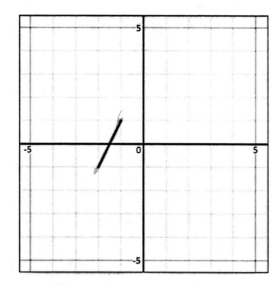

$(-1, 1)$ $(-2, -1)$

Ⓐ $y = 2x + 4$

Ⓑ $y = 2x + 3$

Ⓒ $y = \dfrac{1}{2}x + 3$

Ⓓ $y = -2x + 4$

10. What function includes the line segment graphed below?

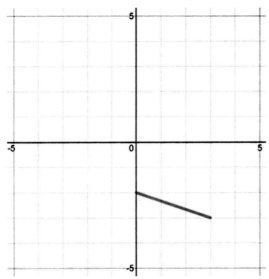

Ⓐ $y = -\dfrac{1}{3}x - 2$

Ⓑ $y = -\dfrac{1}{2}x + 2$

Ⓒ $y = \dfrac{1}{3}x + 2$

Ⓓ $y = \dfrac{1}{2}x - 3$

11. Select the graph that best matches the following description: A car traveling at a constant speed.

Ⓐ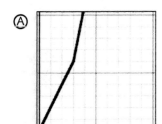

Ⓒ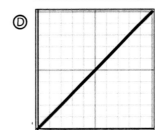

Ⓑ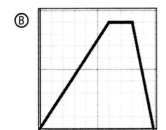

Ⓓ

12. Select the graph that best matches the following description: Going to school in the morning and coming back home in the afternoon.

(A)
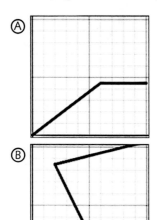

(B)

(C)
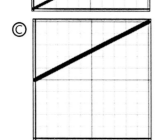

(D)
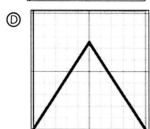

13. Select the graph that best matches the following description: The cost of hiring a plumber per hour including a set fee for coming to the house.

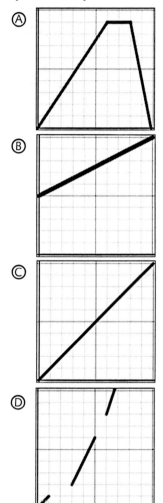

14. Select the graph that best matches the following description: Moving at a steady pace, then stopping.

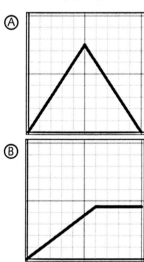

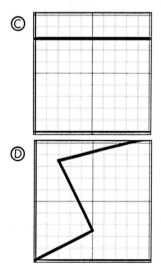

15. Select the graph that best matches the following description: A runner starts with a brisk jog and moves into a sprint.

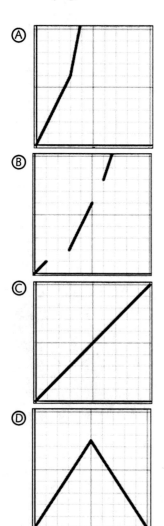

End of Functions

Answer Key and Detailed Explanations
Functions
Functions (8.F.A.1)

Question No.	Answer	Detailed Explanation
1	D	In order to be classified as a function each x must map to one and only one value of y. In the case of the set of ordered pairs {(2, 3), (2, 7), (8, 6)}, the x value of 2 has two different values for y, 3 and 7. Therefore, this set does not classify as a function.
2	B	In the case of <table><tr><td>x</td><td>y</td></tr><tr><td>1</td><td>7</td></tr><tr><td>3</td><td>8</td></tr><tr><td>4</td><td>7</td></tr></table> for each x there is one and only one value for y. Therefore, this table represents a function.
3	A	If y is a function of x, a particular x value CANNOT be associated with two or more values of y.
4	A	In the first graph, for a certain value of x, there are two distinctly different values for y. This is NOT a function.
5	C	In the third graph for certain values of x, there are more than one value for y. This is NOT a function
6	D	{(0, 0), (0, 2), (0, 4)} does not qualify as a function because for x = 0, there is more than one value for y.
7	D	A line drawn through the graph perpendicular to the x-axis will cross the graph one and only one time because for each value of x, there will be 1 and only one value for y.
8	D	In a function, for each value of x, there must be one and only one value for y. We already have (2,1) so cannot have another ordered pair where x=2.
9	D	There is already an assigned y value for x=6, x=5, and x=4, but not for x = 3.
10	D	Only this set, [(-7,10), (7,10), (8,9), (9,10)], represents a function because for each value for x, there exists one and only one value for y.

Comparing Functions (8.F.A.2)

Question No.	Answer	Detailed Explanation
1	A	Only p = 2(n – 5) meets the requirements to subtract 5 from n and then double the result to get p.
2	B	The slope is 2. The only function with a slope of 2 is y = 2x + 5.
3	B	According to the table, when x increases by 1, y increases by 2. Therefore the slope of the line (m) is 2. The only function offered with a slope of 2 is y = 2x - 4.
4	C	Function A has a slope of 1/3 and Function B has a slope of 1/2. Therefore, Function B has a greater slope. To verify, apply the formula: Slope = m =$(y_2-y_1)/(x_2-x_1)$ to each function.
5	C	The slope (m) of Line A = (9-4)/(7 - -1)= 5/8 The slope (m) of Line B = (-3-2)/(-3-5) = (-5)/(-8) = 5/8 Since their slopes are equal, the lines are parallel.
6	A	The slope (m) of Line R = (4+2)/(6+2) = 6/8 = 3/4 The slope (m) of Line S = (4-0)/(0-3) = 4/-3 Since their slopes are negative reciprocals of each other, the lines are perpendicular.
7	A	y = mx + b m = (13-1)/(-4-2) = 12/-6 = -2 y = -2x + b Substitute (x,y) for one point and solve for b. Using (2,1), we get 1 = -2(2) + b 1 = -4 + b 1 + 4 = b 5 = b; so y-intercept is (0,5)
8	C	If you substitute the coordinates of the point (1,9) into each equation, you will find that both equations are satisfied. "The graphs of the functions will intersect at the point (1,9)" is the correct answer.
9	B	Since their slopes, 3 and -1/3, are negative reciprocals of each other, the lines will be perpendicular.

Question No.	Answer	Detailed Explanation
10	A	The slope of line A is 1 and that of line B is 5. Therefore, line B has the steeper slope.

Linear Functions (8.F.A.3)

Question No.	Answer	Detailed Explanation
1	D	The slope of the line through the given points is 1/2. So slope (m) = (11-7)/(k-6) = 1/2 k-6 = 2(11-7) k-6 = 8 k = 14
2	C	In linear equations, none of the variables is of the second power or higher and they can be put into the form of y = mx + b.
3	B	In the function f(x) = x - 3, the coordinates of (3,0) satisfy the equation and it is written in the form of a linear equation.
4	A	The slope is a constant 3/2 between any two points and slope is positive; so y increases as x increases. Therefore, it is increasing linearly.
5	D	Only the last statement, the function is nonlinear because it does not have the same slope between different pairs of points, is true.
6	B	If it is a function, x has to change as y changes, but if that is true, the line cannot be constant for y unless it is in quadrants I and II, or III and IV, or it is the x-axis.
7	C	If the function is linear, the slope has to be the same between any 2 points. Using the first two points, the slope is 2/3. Therefore, using the next two points, the slope must also be 2/3; so (n-3)/(9-3)= 2/3; so 3n-9=18-6 3n=12+9=21 n=7
8	C	(9-3)/(3-0)=(n-9)/(9-3) 6/3=(n-9)/6 2/1 = (n-9)/6 n-9=12 n=21

Question No.	Answer	Detailed Explanation
9	D	If the graph lies in quadrants I and II and is non-negative, it cannot reach quadrant IV, although it could also pass through quadrant III.
10	A	Since the slope is 2, y = 2x + 2 is the only function that could fit.

Linear Function Models (8.F.A.4)

1	D	There is not enough information to say any of the choices MUST be true although it may represent a portion of the graph of a constant function.
2	A	y = mx + 2 0 = 4m + 2 -2 = 4m -2/4 = m -1/2 = m Therefore, m is negative is true.
3	A	y = 2x + b 0 = 2(-5) + b 0 = -10 + b 10 = b Therefore, b is positive is true.
4	D	All that we know is that the y-intercept is (0,0); i.e. b=0.
5	C	m = (5-5)/(9-2)= 0/7=0 y = mx + b 5 = 0 + b 5 = b
6	C	m = (7-5)/(2-3) = 2/-1= -2
7	C	It is linear and the y value is 5 in the case of the two given points. Therefore, it is a constant function.
8	A	The slopes range between 1/2 and 2; so 2 is the greatest slope. The first option is the correct answer.
9	B	b = 18, the height of the bench. The height increases by 3 in. for each block added; so h = 3b + 18
10	C	He starts out $10 in debt so b = -10. At any point, he has earned 5w where w is the number of weeks. Therefore, m = 5w -10.

Question No.	Answer	Detailed Explanation
11	A	A guest must pay $5 admission before any rides; so b=5. He pays $2 per ride so the total paid in rides is 2r. Now add these two amounts and p = 2r + 5
12	D	The slopes range between 1/3 and 1; so 1/3 is the smallest slope. The last choice is the correct answer.
13	B	y = mx + b m = (4-2)/(9-6) = 2/3 Substitute (6,2). 2 = 6(2/3) + b 2 = 4 + b 2-4 =b -2 = b Therefore, y-intercept is (0, -2)
14	D	The point (0,-5) tells us that b = -5. m = (15-0)/(4-1) = 15 / 3 = 5 Substitute into y = mx + b. f(x) = 5x - 5
15	B	Since you pay $20 for the first CD, b = 20. If you purchase x CDs and you already paid $20 for the first one, the remainder would be (x-1) CDs for which you pay $10 each. Therefore, f(x) = 10(x-1) + 20.

Analyzing Functions (8.F.B.5)

1	B	The cost per copy is a function of the number of copies purchased means that the cost per copy changes as the number of copies changes.
2	B	In a positive function, as one variable increases so does the other.
3	C	Her first score was i. She doubled the previous score three times or $2^3 i$.

Question No.	Answer	Detailed Explanation
4	A	The slope is positive in the first two segments. The second segment is steeper than the first, indicating an increase in speed. The flat line represents no rate of change and the time Kevin spends on the bench. Finally, as Kevin returns from the park, his distance from home decreases, which is represented by the final line segment by its negative slope.
5	B	The graph begins with a positve slope, indicating an initial increase in speed. Then, the graph changes to a negative slope at the time when Alison slows down by hitting the brakes.
6	A	Runner A has the steepest slope and is therefore covering more distance in less time than the other runners.
7	A	The speed can be found by calculating the slope of the line. Given the two points (0,0) and (3,6), the slope can be calcuated as (6-0)/(3-0)=2 m/s
8	C	When Karina is relaxing in the tub, the water level is not changing. This is represented by segment C, which has a slope of zero.
9	B	The line segment has a slope of 2. Extend the line to find the y-intercept of 3. Substitute these values into y=mx+b to write the equation y=2x+3.
10	A	The line segment has a slope of -1/3. Extend the line to find the y-intercept of -2. Substitute these values into y=mx+b to write the equation y=(-1/3)x-2.
11	D	The slope of this line remains constant, representing a car traveling at a constant speed.
12	D	The distance from home increases over time on the way to school and decreases on the way back home.
13	B	The set fee is represented as fixed on the y-intercept while the slope of the line represents the constant slope of the hourly rate.

Question No.	Answer	Detailed Explanation
14	B	The positive slope of the first segment represents a rate of change, while the horizontal line with zero slope represent stopping because there is no change over time.
15	A	The slope is positive in the first two segments. The second segment is steeper than the first, indicating an increase in speed, from jog to sprint in this case.

Geometry

Transformations of Points & Lines (8.G.A.1.A)

1. The point (4, 3) is rotated 90° clockwise about the origin. What are the coordinates of the resulting point?

   Ⓐ (-3, 4)
   Ⓑ (-4, 3)
   Ⓒ (4, -3)
   Ⓓ (3, -4)

2. A line segment has a length of 9 units. After a certain transformation is applied to the segment, the new segment has a length of 9 units. What was the transformation?

   Ⓐ A rotation
   Ⓑ A reflection
   Ⓒ A translation
   Ⓓ All choices are correct.

3. The point (2, 4) is rotated 180° clockwise about the origin. What are the coordinates of the resulting point?

   Ⓐ (-2, -4)
   Ⓑ (-2, 4)
   Ⓒ (2, -4)
   Ⓓ (2, 4)

4. Two points are located in the (x, y) plane. After a certain transformation is applied to both points, the two new points end up on the opposite side of the y-axis. What was the transformation?

   Ⓐ A rotation
   Ⓑ A reflection
   Ⓒ A translation
   Ⓓ A dilation

5. A certain transformation is applied to a line segment. The new segment shifted to the left within the coordinate plane. What was the transformation?

   Ⓐ A rotation
   Ⓑ A reflection
   Ⓒ A translation
   Ⓓ It cannot be determined.

6. A line segment with end points (1, 1) and (5, 5) is moved and the new end points are now (1, 5) and (5,1). Which transformation took place?

Ⓐ reflection
Ⓑ rotation
Ⓒ translation
Ⓓ dilation

7. A certain transformation moves a line segment as follows: A (2, 1) moves to A' (2, -1) and B (5, 3) to B' (5, -3).
Name this transformation.

Ⓐ Rotation
Ⓑ Translation
Ⓒ Reflection
Ⓓ Dilation

8. After a certain transformation is applied to point (x, y), it moves to (y, -x).
Name the transformation.

Ⓐ Rotation
Ⓑ Translation
Ⓒ Reflection
Ⓓ Dilation

9. A transformation moves the point (0, y) to a new location at (0, -y).
Name this transformation.

Ⓐ Rotation
Ⓑ Translation
Ⓒ Reflection
Ⓓ It could be any one of the three listed.

10. (x,y) is in Quadrant 1. Reflection across the y-axis would move it to a point with the following coordinates.

Ⓐ (x, y)
Ⓑ (-x, -y)
Ⓒ (x, -y)
Ⓓ (-x, y)

Transformations of Angles (8.G.A.1.B)

1. △**ABC** is reflected across the x-axis.
    What two angles are equivalent?

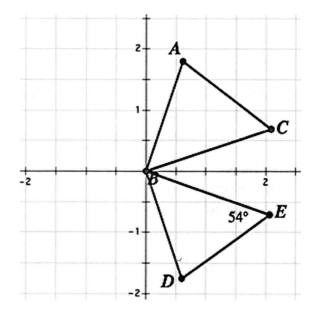

    Ⓐ ∠ A and ∠ C
    Ⓑ ∠ A and ∠ E
    Ⓒ ∠ C and ∠ D
    Ⓓ ∠ C and ∠ E

2. △**ABC** is rotated 90°.
    What two angles are equivalent?

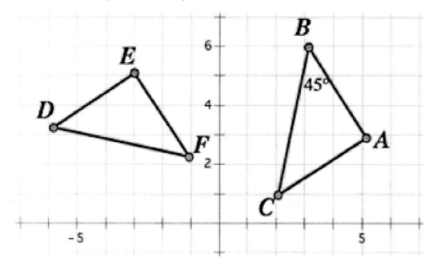

    Ⓐ ∠A and ∠C
    Ⓑ ∠B and ∠E
    Ⓒ ∠C and ∠D
    Ⓓ ∠C and ∠F

3. What rigid transformation should be used to prove ∠ABC ≅ ∠XYZ ?

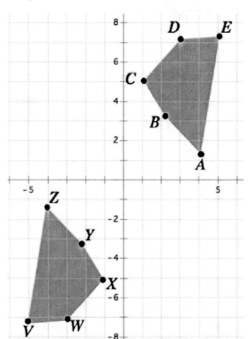

ⒶReflection
ⒷRotation
ⒸTranslation
ⒹNone of the above

4. What rigid transformation should be used to prove ∠ABC ≅ ∠XYZ?

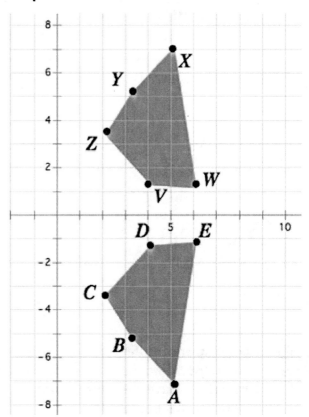

ⒶReflection
ⒷRotation
ⒸTranslation
ⒹNone of the above

5. △**ABC** is rotated 90°.
    What two angles are equivalent?

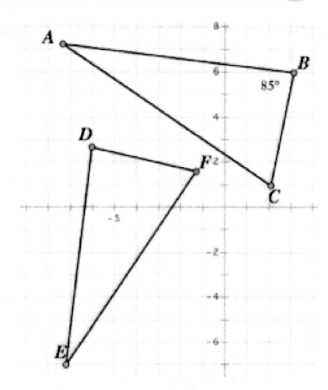

ⓐ ∠ A and ∠ F
ⓑ ∠ B and ∠ D
ⓒ ∠ B and ∠ F
ⓓ ∠ C and ∠ D

6. **If all the triangles below are the result of one or more rigid transformations, which of the following MUST be true?**

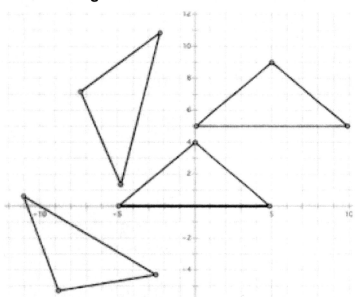

ⓐ All corresponding line segments are congruent.
ⓑ All corresponding angles are congruent.
ⓒ All triangles have the same area.
ⓓ A,B, and C are all correct.

7. **What rigid transformation should be used to prove** ∠**ABC** ≅ ∠**DEF**?

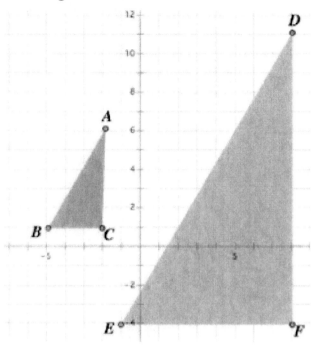

Ⓐ Reflection
Ⓑ Rotation
Ⓒ Translation
Ⓓ None of the above

8. **A company is looking to design a new logo, which consists only of transformations of the angle below:**

Which logo meets the company's demand?

Ⓐ

Ⓑ
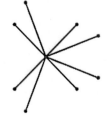

©

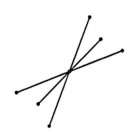

ⓓ **All of the above**

9. The angle ∠AOB is 45° and has been rotated 120° around point C. What is the measure of the new angle ∠XYZ?

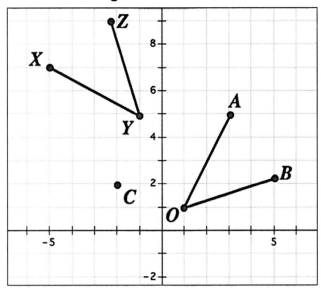

Ⓐ 30°
Ⓑ 45°
Ⓒ 90°
Ⓓ 120°

10. Find the measure of ∠ABC

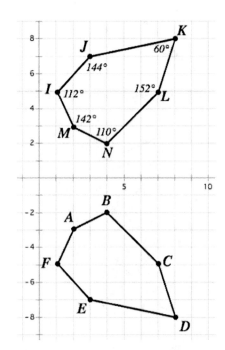

Ⓐ 110°
Ⓑ 112°
Ⓒ 142°
Ⓓ 144°

Transformations of Parallel Lines (8.G.A.1.C)

1. Two parallel line segments move from Quadrant One to Quadrant Four. Their slopes do not change. What transformation has taken place?

   Ⓐ Reflection
   Ⓑ Translation
   Ⓒ Dilation
   Ⓓ This is not a transformation.

2. Two parallel line segments move from Quadrant One to Quadrant Four. Their slopes change from a positive slope to a negative slope. What transformation has taken place?

   Ⓐ Reflection
   Ⓑ Rotation
   Ⓒ Translation
   Ⓓ It could be either a rotation or a reflection.

3. Two parallel line segments move from Quadrant One to Quadrant Two. Their slopes change from a negative slope to a positive slope. What transformation has taken place?

   Ⓐ Reflection
   Ⓑ Rotation
   Ⓒ Translation
   Ⓓ It could be either a rotation or a reflection.

4. Line *L* is translated along segement $\overline{AB}$ to create line *L'* . Will *L* and *L'* ever intersect?

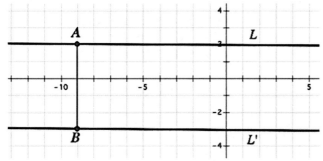

   Ⓐ Yes, line *L'* is now the same as *L*.
   Ⓑ Yes, parallel lines always eventually intersect.
   Ⓒ No, every point on *L'* will always have a corresponding point the distance of $\overline{AB}$ away from *L'*.
   Ⓓ No, the translation along $\overline{AB}$ does not change the slope from *L* to *L'*, and lines with the same slope never intersect.

5. Line *L* is translated along segement $\overline{AB}$ to create line *L'*. Will *L* and *L'* ever intersect?

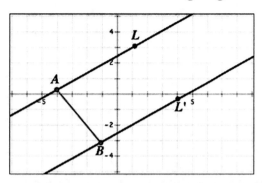

   Ⓐ Yes, line *L'* is now the same as *L*.
   Ⓑ Yes, parallel lines always eventually intersect.
   Ⓒ No, every point on *L'* will always have a corresponding point on *L'*.
   Ⓓ No, the translation along $\overline{AB}$ does not change the slope from *L* to *L'*, and lines with the same slope never intersect.

6. Line *L* is translated along ray $\overline{AC}$ to create line *L'*. What do you know about the relationship between line *L* and *L'*?

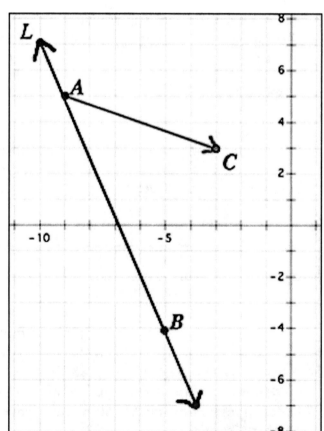

   Ⓐ The lines intersect at least once.
   Ⓑ The lines are exactly the same.
   Ⓒ The lines are parallel.
   Ⓓ None of the above.

7. How many lines can be drawn through point *C* that are parallel to line *L*?

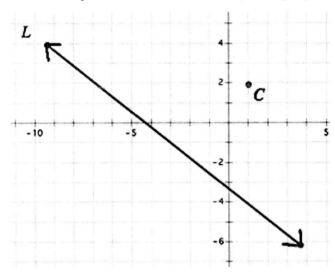

Ⓐ None
Ⓑ One
Ⓒ Two
Ⓓ Infinitely many

8. Figure *ABCD* was rotated around the origin to create *WXYZ*. Prove $\overline{XY}$ and $\overline{WZ}$ are parallel.

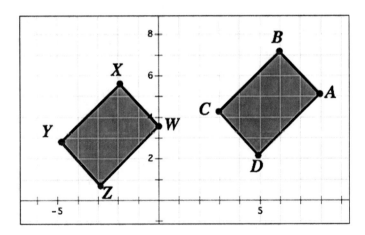

Ⓐ The sides of all rectangles are parallel.

Ⓑ Since *ABCD* is a rectangle, $\overline{AD} \parallel \overline{BC}$. Translations map parallel lines to parallel lines, so $\overline{XY} \parallel \overline{WZ}$.

Ⓒ It cannot be proven because $\overline{XY}$ and $\overline{WZ}$ are perpendicular.

Ⓓ It cannot be proven because the angle of rotation is not given.

9. **The two parallel lines shown are rotated 180° abound the origin. What is the result of this transformation?**

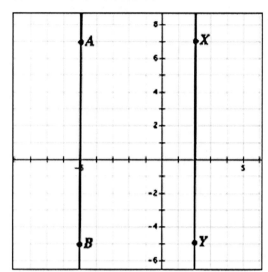

Ⓐ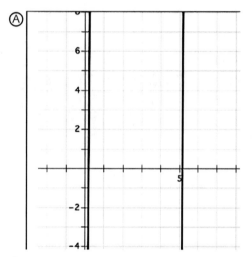

Ⓒ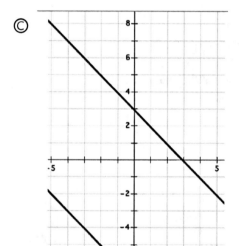

Ⓑ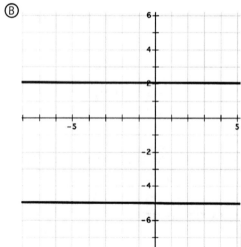

Ⓓ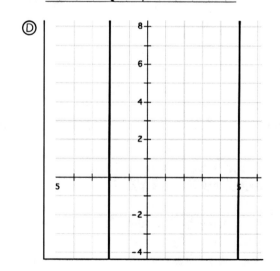

10. Figure *DEF* is the result of 180° rotation around the origin of figure *ABC*. Prove $\overline{AB}$ and $\overline{DE}$ are parallel.

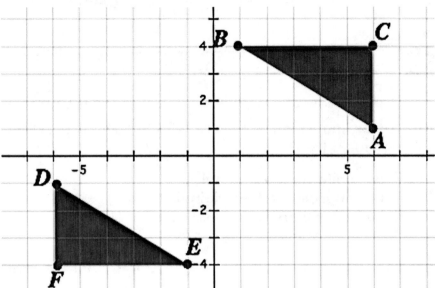

Ⓐ A 180° rotation of a given segment always maps to a parallel segment. Therefore $\overline{AB} \mid\mid \overline{DE}$.

Ⓑ Corresponding sides of triangles are always parallel. Therefore, $\overline{AB} \mid\mid \overline{DE}$.

Ⓒ It cannot be proven because $\overline{XY}$ and $\overline{WZ}$ are perpendicular.

Ⓓ It cannot be proven due to the angle of rotation is not given.

Transformations of Congruency (8.G.A.2)

1. Which of the following examples best represents congruency in nature?

    Ⓐ A mother bear and her cub.
    Ⓑ The wings of a butterfly.
    Ⓒ The tomatoes picked from my garden.
    Ⓓ The clouds in the sky.

2. If triangle ABC is drawn on a coordinate plane and then reflected over the vertical axis, which of the following statements is true?

    Ⓐ The reflected triangle will be similar to the original.
    Ⓑ The reflected triangle will be congruent to the original.
    Ⓒ The reflected triangle will be larger than the original.
    Ⓓ The reflected triangle will be smaller than the original.

3. What transformation was applied to the object in quadrant 2 to render the results in the graph below?

    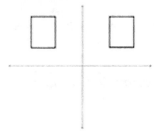

    Ⓐ reflection
    Ⓑ rotation
    Ⓒ translation
    Ⓓ not enough information

4. What transformation was applied to the object in quadrant 2 to render the results in the graph below?

    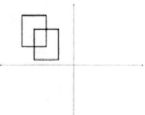

    Ⓐ reflection
    Ⓑ rotation
    Ⓒ translation
    Ⓓ not enough information

5. What is NOT true about the graph below?

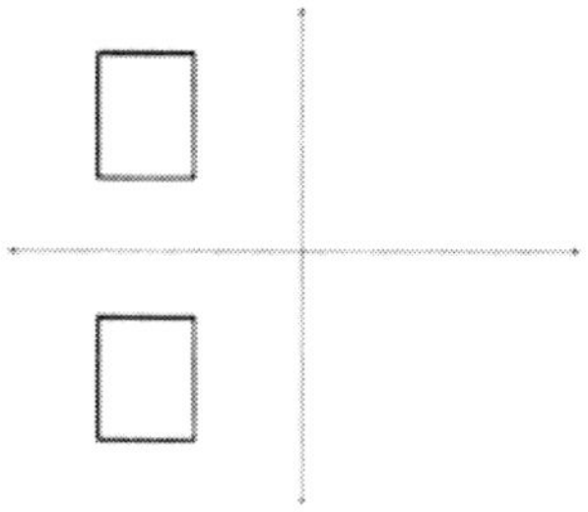

Ⓐ The object in quadrant 3 could not be a reflection of the object in quadrant 2.
Ⓑ The object in quadrant 3 could be a translation of the object in quadrant 2.
Ⓒ The two objects are congruent.
Ⓓ The object in quadrant 3 could not be a dilation of the object in quadrant 2.

6. Finish the statement. Two congruent objects _____.

Ⓐ have the same dimensions.
Ⓑ have different measured angles.
Ⓒ are not the same shape.
Ⓓ only apply to two-dimensional objects.

7. What transformations can be applied to an object to create a congruent object?

Ⓐ all transformations
Ⓑ dilation and rotation
Ⓒ translation and dilation
Ⓓ reflection, translation, and rotation

8. A figure formed by rotation followed by reflection of an original triangle will be _____.

Ⓐ similar to the original.
Ⓑ congruent to the original.
Ⓒ smaller than the original.
Ⓓ larger than the original.

9. Which of the following is NOT a characteristic of congruent triangles?

Ⓐ They have three pairs of congruent sides.
Ⓑ They have three pairs of congruent angles.
Ⓒ Their areas are equal.
Ⓓ They have four pairs of proportional sides

10. **Which of the following letters looks the same after a reflection followed by a 180°**
    rotation.

Ⓐ **P**
Ⓑ **O**
Ⓒ **F**
Ⓓ **None of the above.**

Analyzing Transformations (8.G.A.3)

1. Right triangle ABC has the following side lengths: AB = 6, BC = 8, AC = 10. After dilation by 1/2, which of the following could be the length of AC?

   Ⓐ 4
   Ⓑ 5
   Ⓒ 10
   Ⓓ 20

2. Which of the following statements is NOT true about transformations?

   Ⓐ A rotation could be described as a turn.
   Ⓑ A reflection could be described as a flip.
   Ⓒ A translation could be described as a slide.
   Ⓓ A dilation could be described as changing direction.

3. Which of the following consists of a size transformation?

   Ⓐ Rotation
   Ⓑ Dilation
   Ⓒ Translation
   Ⓓ Reflection

4. Which of the following transformations could transform triangle A to triangle B?

    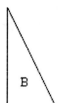

   Ⓐ Rotation
   Ⓑ Reflection
   Ⓒ Translation
   Ⓓ Dilation

5. Which of the following transformations could transform triangle A to triangle B?

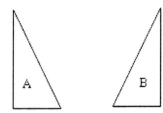

Ⓐ Rotation
Ⓑ Reflection
Ⓒ Translation
Ⓓ Dilation

6. Which of the following transformations could transform triangle A to triangle B?

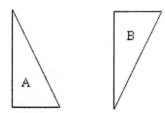

Ⓐ Rotation
Ⓑ Reflection
Ⓒ Translation
Ⓓ None of the above

7. Which of the following sequences of transformations could transform triangle A to triangle B?

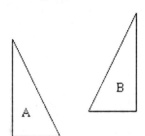

Ⓐ A reflection followed by a translation
Ⓑ A reflection followed by another reflection
Ⓒ A rotation followed by a translation
Ⓓ A dilation followed by a translation

8. Which of the following transformations does NOT preserve congruency?

    Ⓐ Rotation
    Ⓑ Translation
    Ⓒ Reflection
    Ⓓ Dilation

9. Consider the triangle with vertices (1, 0), (2, 5) and (-1, 5). Find the vertices of the new triangle after a reflection over the vertical axis followed by a reflection over the horizontal axis.

    Ⓐ (-1, 0), (1, 5) and (-2, 5)
    Ⓑ (-1, 0), (-2, -5) and (1, -5)
    Ⓒ (1, 0), (-2, 5) and (1, 5)
    Ⓓ (1, 0), (-2, -5) and (1, -5)

10. Translate the triangle with vertices (1, 0), (2, 5), and (-1, 5), 3 units to the left. Which of the following ordered pairs represent the vertices of the new triangle?

    Ⓐ (-2, 0), (-4, 5) and (-1, 5)
    Ⓑ (4, 0), (5, 5) and (2, 5)
    Ⓒ (-1, 2), (2, 2) and (1, -3)
    Ⓓ (-1, 8), (2, 8) and (1, 3)

Transformations & Similarity (8.G.A.4)

1. Which of the following transformations could transform triangle A to triangle B?

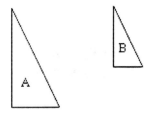

   Ⓐ Rotation
   Ⓑ Reflection
   Ⓒ Translation
   Ⓓ Dilation

2. What transformations have been applied to the large object to render the results in the graph below?

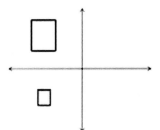

   Ⓐ reflection and dilation
   Ⓑ rotation and translation
   Ⓒ rotation and dilation
   Ⓓ translation and reflection

3. What transformations have been applied to the large object to render the results in the graph below?

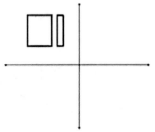

   Ⓐ no transformation
   Ⓑ reflection and dilation
   Ⓒ translation and dilation
   Ⓓ rotation and dilation

4. Which graph represents reflection over an axis and dilation?

Ⓐ

Ⓑ

Ⓒ

Ⓓ None of the above.

5. What transformation is necessary to have two similar, but not necessarily congruent, objects?

Ⓐ Rotation
Ⓑ Translation
Ⓒ Dilation
Ⓓ Reflection

6. Finish this statement. Two similar objects _____.

Ⓐ have proportional dimensions.
Ⓑ are always congruent.
Ⓒ have different measured angles.
Ⓓ can be different shapes.

7. If a point, P, on a coordinate plane moves from (9, 3) to P' (3, -9) and then to P'' (0, -9), what transformations have been applied?

Ⓐ Dilation followed by Translation
Ⓑ Translation followed by Dilation
Ⓒ Rotation followed by Translation
Ⓓ Translation followed by Rotation

8. Consider Triangle ABC, where AB = 5, BC = 3, and AC = 6, and Triangle WXY, where WX = 10, XY = 6, and WY = 12.
   Assume ABC is similar to WXY.
   Which of the following represents the ratio of similarity?

   Ⓐ 1 : 2
   Ⓑ 5 : 6
   Ⓒ 3 : 10
   Ⓓ 6 : 20

9. Rectangle A is 1 unit by 2 units.
   Rectangle B is 2 units by 3 units.
   Rectangle C is 2 units by 4 units.
   Rectangle D is 3 units by 6 units.
   Which rectangle is not similar to the other three rectangles?

   Ⓐ A
   Ⓑ B
   Ⓒ C
   Ⓓ D

10. If triangle ABC is similar to triangle WXY and AB = 9, BC = 7, AC = 14, WX = 27, and XY = 21.
    Find WY.

   Ⓐ 44
   Ⓑ 43
   Ⓒ 42
   Ⓓ 41

Interior & Exterior Angles in Geometric Figures (8.G.A.5)

1. What term describes a pair of angles formed by the intersection of two straight lines that share a common vertex but do not share any common sides?

   Ⓐ Supplementary Angles
   Ⓑ Complementary Angles
   Ⓒ Horizontal Angles
   Ⓓ Vertical Angles

2. If a triangle has two angles with measures that add up to 100 degrees, what must the measure of the third angle be?

   Ⓐ 180 degrees
   Ⓑ 100 degrees
   Ⓒ 80 degrees
   Ⓓ 45 degrees

3.

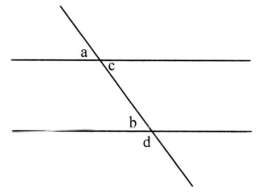

   The figure shows two parallel lines intersected by a third line. If a = 55°, what is the value of b?

   Ⓐ 35°
   Ⓑ 45°
   Ⓒ 55°
   Ⓓ 125°

4.

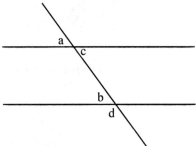

The figure shows two parallel lines intersected by a third line. If b = 60°, what is the value of c?

Ⓐ 30°
Ⓑ 60°
Ⓒ 90°
Ⓓ 120°

5.

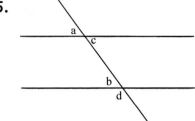

The figure shows two parallel lines intersected by a third line. If d = 130°, what is the value of a?

Ⓐ 30°
Ⓑ 40°
Ⓒ 50°
Ⓓ 130°

6.

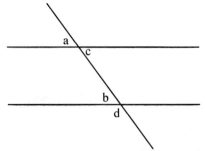

The figure shows two parallel lines intersected by a third line. Which of the following angles are equal in measure?

Ⓐ a and b only
Ⓑ a and c only
Ⓒ b and c only
Ⓓ a, b, and c

7. Two angles in a triangle measure 65° each. What is the measure of the third angle in the triangle?

Ⓐ 25°
Ⓑ 50°
Ⓒ 65°
Ⓓ 130°

8.

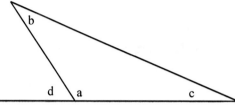

If b = 40° and c = 30°, what is the measure of d?

Ⓐ 35°
Ⓑ 70°
Ⓒ 110°
Ⓓ 145°

9. In right triangle ABC, Angle C is the right angle. Angle A measures 70°. Find the measure of the exterior angle at angle C.

Ⓐ 180°
Ⓑ 90°
Ⓒ 110°
Ⓓ 160°

10. If two parallel lines are cut by a transversal, the alternate interior angles are _____.

Ⓐ supplementary
Ⓑ complementary
Ⓒ equal in measure
Ⓓ none of the above

Verifying the Pythagorean Theorem (8.G.B.6)

1. **Which of the following could be the lengths of the sides of a right triangle?**

   Ⓐ 1, 2, 3
   Ⓑ 2, 3, 4
   Ⓒ 3, 4, 5
   Ⓓ 4, 5, 6

2. **A triangle has sides 8 cm long and 15 cm long, with a 90° angle between them. What is the length of the third side?**

   Ⓐ 7 cm
   Ⓑ 17 cm
   Ⓒ 23 cm
   Ⓓ 289 cm

3. **Find the value of c, rounded to the nearest tenth.**

   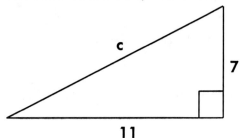

   Ⓐ 8.5
   Ⓑ 8.8
   Ⓒ 13.0
   Ⓓ 19.3

4. **A square has sides 5 inches long. What is the approximate length of a diagonal of the square?**

   Ⓐ 5 inches
   Ⓑ 6 inches
   Ⓒ 7 inches
   Ⓓ 8 inches

5. Which of the following INCORRECTLY completes this statement of the Pythagorean theorem?
   In a right triangle with legs of lengths a and b and hypotenuse of length c, ...

   Ⓐ $a^2 + b^2 = c^2$
   Ⓑ $c^2 - a^2 = b^2$
   Ⓒ $c^2 - b^2 = a^2$
   Ⓓ $a^2 + c^2 = b^2$

6. A Pythagorean triple is a set of three positive integers a, b, and c that satisfy the equation $a^2 + b^2 = c^2$. Which of the following is a Pythagorean triple?

   Ⓐ a = 3, b = 6, c = 9
   Ⓑ a = 6, b = 9, c = 12
   Ⓒ a = 9, b = 12, c = 15
   Ⓓ a = 12, b = 15, c = 18

7. If an isosceles right triangle has legs of 4 inches each, find the length of the hypotenuse.

   Ⓐ Approximately 6 in.
   Ⓑ Approximately 5 in.
   Ⓒ Approximately 4 in.
   Ⓓ Approximately 3 in.

8. In triangle ABC, angle C = 90°, AC = 4 and AB = 10. Find BC to the nearest tenth.

   Ⓐ 9.5
   Ⓑ 9.2
   Ⓒ 8.9
   Ⓓ 8.5

9. In triangle ABC, angle C = 90°, AB = 35, and BC = 28. Find AC.

   Ⓐ 23
   Ⓑ 22
   Ⓒ 21
   Ⓓ 20

10. The diagonal of a square is 25. Find the approximate side lengths.

   Ⓐ 15
   Ⓑ 16
   Ⓒ 17
   Ⓓ 18

1. The bottom of a 17-foot ladder is placed on level ground 8 feet from the side of a house as shown in the figure below. Find the vertical height at which the top of the ladder touches the side of the house.

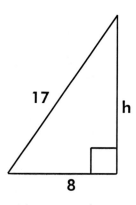

   Ⓐ h = 9 feet
   Ⓑ h = 12 feet
   Ⓒ h = 15 feet
   Ⓓ h = 18 feet

2. Which of the following equations could be used to find the value of w?

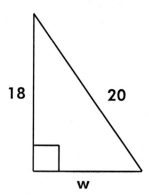

   Ⓐ $w^2 + 18^2 = 20^2$
   Ⓑ $18^2 - w^2 = 20^2$
   Ⓒ $20^2 + 18^2 = w^2$
   Ⓓ $w + 18 = 20$

3. John has a chest where he keeps his antiques. What is the measure of the diagonal (d) of John's chest with the height (c) = 3ft, width (b) = 3ft, and length (a) = 5ft.?

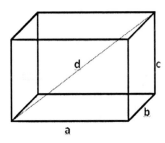

   Ⓐ √42 ft²
   Ⓑ √43 ft²
   Ⓒ √34 ft
   Ⓓ √43 ft

4. Mary has a lawn that has a width (a) of 30 yards, and a length (b) of 40 yards. What is the measurement of the diagonal (c) of the lawn?

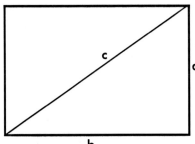

   Ⓐ 50 yards
   Ⓑ 49 yards
   Ⓒ 51 yards
   Ⓓ 50 yards²

5. A construction company needed to build a sign with the width (a) of 9 ft, and a length (b) of 20 ft. What will be the approximate measurement of the diagonal (c) of the sign?

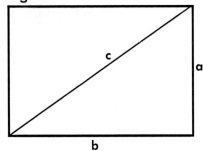

   Ⓐ 23 ft
   Ⓑ 22 ft
   Ⓒ 20 ft
   Ⓓ 19 ft

6. An unofficial baseball diamond is measured to be 50 yards wide. What is the approximate measurement of one side (a) of the diamond?

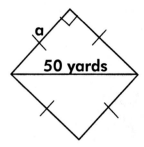

Ⓐ 34 ft
Ⓑ 35 yards
Ⓒ 35 ft
Ⓓ 36 yards

7. The neighborhood swimming pool is 20 ft wide and 30 ft long. What is the approximate measurement of the diagonal (d) of the base of the pool?

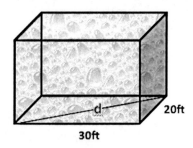

Ⓐ 36 ft
Ⓑ 35 ft
Ⓒ 34 ft
Ⓓ 34 yards

8. To get to her friend's house, a student must walk 20 feet to the corner of their streets, turn left and walk 15 feet to her friend's house.
How much shorter would it be if she could cut across a neighbor's yard and walk a straight line from her house to her friend's house?

Ⓐ 5 feet shorter
Ⓑ 10 feet shorter
Ⓒ 25 feet shorter
Ⓓ 35 feet shorter

9. Your school principal wants the custodian to put a new flag up on the flagpole. If the flagpole is 40 feet tall and they have a 50 foot ladder, approximately how far from the base of the pole can he place the base of his ladder in order to accomplish the task?

Ⓐ Up to 10 ft away
Ⓑ Up to 20 ft away
Ⓒ Up to 30 ft away
Ⓓ Up to 40 ft away

10. Your kite is at the end of a 50 ft string. You are 25 ft from the outside wall of a building that you know makes a right angle with the ground.
If the kite flyer is 5 foot tall, approximately how high is the kite?

Ⓐ Approximately 39 ft high
Ⓑ Approximately 40 ft high
Ⓒ Approximately 42 ft high
Ⓓ Approximately 48 ft high

Pythagorean Theorem & Coordinate System (8.G.B.8)

1. A robot begins at point A, travels 4 meters west, then turns and travels 7 meters south, reaching point B. What is the approximate straight-line distance between points A and B?

    (A) 8 meters
    (B) 9 meters
    (C) 10 meters
    (D) 11 meters

2. What is the distance between the points (1, 3) and (9, 9)?

    (A) 6 units
    (B) 8 units
    (C) 10 units
    (D) 12 units

3. Find the distance between A (2, 7) and B (-2, -7).

    (A) 14.0
    (B) 14.6
    (C) 18.0
    (D) 13.4

4. Find the distance between P (5, 3) and the origin (0, 0).

    (A) 4.0
    (B) 5.1
    (C) 5.8
    (D) 8.0

5. Find the distance between the points A (11, 12) and B (7, 8).

    (A) 4.0
    (B) 5.3
    (C) 5.7
    (D) 8.0

6. Is it closer to go from A (4, 6) to B (2, -4) or A to C (-5, 2)?

    (A) A to B
    (B) A to C
    (C) Neither, they are both the same distance.
    (D) Not enough information.

7. Paul lives 50 yards east and 40 yards south of his friend, Larry. If he wants to shorten his walk by walking in a straight line from his home to Larry's, how far will he walk?

Ⓐ 64 yards
Ⓑ 62 yards
Ⓒ 60 yards
Ⓓ 58 yards

8. In a coordinate plane, a point P (7,8) is rotated 90° clockwise around the origin and then reflected across the vertical axis. Find the distance between the original point and the final point.

Ⓐ 14 units
Ⓑ 21 units
Ⓒ 24 units
Ⓓ None of the above

9. You are using a coordinate plane to sketch out a plan for your vegetable garden. Your garden will be a rectangle 15 ft wide and 20 ft long. You want a square in the center with 3 ft sides to be reserved for flowers. If the garden is plotted on the coordinate plane with the southwest corner at the origin, what are the coordinates of the center of the flower garden?

Ⓐ (15, 10)
Ⓑ (7.5, 10)
Ⓒ (-15, 10)
Ⓓ (-7.5, 10)

10. You are using a coordinate plane to sketch out a plan for a vegetable garden. The garden will be a rectangle 15 feet wide and 20 feet long. You want a square in the center with 3 foot sides to be reserved for flowers. The garden is plotted in the coordinate grid so that the southwest corner is placed at the origin. Find the length of the diagonal of the flower bed.

Ⓐ 4.0 feet
Ⓑ 4.2 feet
Ⓒ 5.0 feet
Ⓓ 5.2 feet

Finding Volume: Cone, Cylinder, & Sphere (8.G.C.9)

1. **What is the volume of a sphere with a radius of 6?**

    (A) 72π
    (B) 144π
    (C) 216π
    (D) 288π

2. **A cone has a height of 9 and a base whose radius is 4. Find the volume of the cone.**

    (A) 18π
    (B) 36π
    (C) 48π
    (D) 72π

3. **What is the volume of a cylinder with a radius of 5 and a height of 3?**

    (A) 30π
    (B) 45π
    (C) 75π
    (D) 120π

4. **A round ball has a diameter of 10 inches. What is the approximate volume of the ball?**

    (A) 130 cubic inches
    (B) 260 cubic inches
    (C) 390 cubic inches
    (D) 520 cubic inches

5. **A cylindrical can has a height of 5 inches and a diameter of 4 inches. What is the approximate volume of the can?**

    (A) 20 cubic inches
    (B) 100 cubic inches
    (C) 200 cubic inches
    (D) 314 cubic inches

6. **A cylinder and a cone have the same radius and the same volume. How do the heights compare?**

    (A) The height of the cylinder is 3 times the height of the cone.
    (B) The height of the cylinder is 2 times the height of the cone.
    (C) The height of the cone is 2 times the height of the cylinder.
    (D) The height of the cone is 3 times the height of the cylinder.

7. **Which of the following has the greatest volume?**

   Ⓐ A sphere with a radius of 2 cm
   Ⓑ A cylinder with a height of 2 cm and a radius of 2 cm
   Ⓒ A cone with a height of 4 cm and a radius of 3 cm
   Ⓓ All three volumes are equal

8. **Which of the following has the greatest volume?**

   Ⓐ A sphere with a radius of 3 cm
   Ⓑ A cylinder with a height of 4 cm and a radius of 3 cm
   Ⓒ A cone with a height of 3 cm and a radius of 6 cm
   Ⓓ All three volumes are equal

9. **If a sphere and a cone have the same radii and the cone has a height of 4, find the ratio of the volume of the sphere to the volume of the cone.**

   Ⓐ r : 1
   Ⓑ 1 : r
   Ⓒ r : 4
   Ⓓ 4 : r

10. **Which of the following describes the relationship between the volumes of a cone and cylinder with the same radii and the same height.**

    Ⓐ The volume of the cylinder is three times that of the cone.
    Ⓑ The volume of the cylinder is 1/3 that of the cone.
    Ⓒ The volume of the cylinder is 4/3 that of the cone.
    Ⓓ None of the above.

End of Geometry

Answer Key and Detailed Explanations
Geometry
Transformations of Points & Lines (8.G.A.1.A)

Question No.	Answer	Detailed Explanation
1	D	If you rotate (4,3) 90° clockwise, it will move from quadrant I to quadrant IV. x becomes 3 and y becomes -4.
2	D	There is not enough information provided. All three of the transformations listed would keep the length of the segment the same.
3	A	x becomes -x and y becomes -y as the point rotates from quadrant I to quadrant III.
4	B	This is a reflection across the y-axis.
5	C	A simple slide to the left within the coordinate plane is an example of a translation.
6	B	This is a rotation, since the segment is still the same length, it is just oriented differently.
7	C	This is a reflection in the x-axis, since the x-coordinates are remaining unchanged and the y-coordinates are switching sign.
8	A	This would be a rotation through 90° clockwise.
9	D	This is either a reflection in the x-axis, a translation downward, or a 180° rotation about the origin.
10	D	Reflection across the y-axis will change the sign of x but keep the same y value.

Transformations of Angles (8.G.A.1.B)

1	D	Corresponding angles are congruent under rigid transformations. Since triangle BDE is a reflection image of triangle ABC, and angle C corresponds to angle E, the two angles are equal in measure.
2	D	Corresponding angles are congruent under rigid transformations. Since triangle DEF is a rotated image of triangle ABC, and angle C corresponds to angle F, the two angles are equal in measure.
3	B	Since figure VWXYZ is a rotated image of figure ABCDE, this transformation will prove angle ABC is congruent to angle XYZ.

Question No.	Answer	Detailed Explanation
4	A	Since figure VWXYZ is a reflected image of figure ABCDE, this transformation will prove angle ABC is congruent to angle XYZ.
5	B	Corresponding angles are congruent under rigid transformations. Since triangle DEF is a rotated image of triangle ABC, and angle B corresponds to angle D, the two angles are equal in measure.
6	D	All figures that undergo rigid transformations maintain their congruency, which includes side measures, angles, and areas. Thus, D is the correct answer.
7	C	Though the two triangles are not congruent, angle ABC can undergo translation to overlap and prove congruency with angle DEF.
8	D	All of the logos are a result of only reflection, rotation, or translation. Therefore, all the logos would be acceptable.
9	B	All figures that undergo rigid transformations have corresponding angles that maintain congruency. Since angle AOB corresponds to angle XYZ, the two are congruent and therefore both 45 degrees.
10	A	Corresponding angles are congruent under rigid transformations. Since figure ABCDEF is a rotated image of figure IJKLMN, and angle ABC corresponds to angle MNL, the two angles are congruent and therefore both measure 110 degrees.

Transformations of Parallel Lines (8.G.A.1.C)

Question No.	Answer	Detailed Explanation
1	B	This is a translation, since the two segments are still oriented in the same way. They were just shifted downward in the coordinate grid.
2	D	This transformation is not a translation, since the line segments are now directed differently. It could be a reflection or a rotation, since both would change the way the segments are facing.
3	D	This transformation is not a translation, since the line segments are now directed differently. It could be a reflection or a rotation, since both would change the way the segments are facing.
4	D	When a line is translated along a vector, the resulting line is parallel.
5	D	When a line is translated along a vector, the resulting line is parallel.
6	C	When a line is translated along a vector, the resulting line is parallel.
7	B	Given a line and a point, there is one and only one line that can be drawn through the point that is parallel to the original line.
8	B	Since ABCD is a rectangle, its opposite sides are parallel by definintion. Translations always map parallel lines to parallel lines, therefore $\overline{XY} \parallel \overline{WZ}$
9	D	Vertical parallel lines rotated 180 degrees will result in another set of parallel vertical lines. Option A is not the answer because it is the result of a translation, not a rotation.
10	A	A 180 degree rotation will always map to a parallel segment.

Transformations of Congruency (8.G.A.2)

Question No.	Answer	Detailed Explanation
1	B	Of the choices given, the wings of a butterfly are nearly always congruent; i.e. the same shape and the same size.
2	B	Reflection preserves congruence.
3	D	We can't tell if it was a reflection in the y-axis or a translation.
4	C	It shows a translation.
5	A	The object in quadrant III could be a reflection of the object in quadrant II. Therefore, the words could not makes the statement false.
6	A	Two congruent objects are identical and consequently have the same dimensions.
7	D	Only dilation does not preserve congruency.
8	B	A figure formed by rotation, reflection or translation or any combination of these three transformations will be congruent to the original.
9	D	Triangles are congruent if they have congruent sides, congruent angles, and equal areas.
10	B	Because the letter O doesn't have a right and left that are different and a top and bottom that are different, it looks the same after this sequence of transformations.

Analyzing Transformations (8.G.A.3)

Question No.	Answer	Detailed Explanation
1	B	If the triangle is dilated by 1/2, each of its sides will become 1/2 of its original length.
2	D	A dilation has nothing to do with direction. The figure is shrunk or enlarged without changing the shape.
3	B	Size is preserved in rotation, reflection, and translation, but not in dilation.
4	C	A translation slides a figure or point without any alterations in size or shape.
5	B	Triangle A has been reflected to produce Triangle B.
6	D	Triangle A cannot be transformed into Triangle B through just one transformation.
7	A	Triangle A has been reflected and translated to produce Triangle B.
8	D	Dilation does not preserve congruency.
9	B	A point (x, y) reflected in the vertical axis becomes (-x, y). A point (- x, y) reflected in the horizontal axis would become (-x, -y).
10	A	When translating a point (x, y) 3 units to the left, it will be located at (x-3, y).

Transformations & Similarity (8.G.A.4)

1	D	A dilation could transform Triangle A to Triangle B because it shrinks or enlarges a figure.
2	A	The large object has been dilated and reflected to produce the small object.
3	A	Since the shape of the object has changed, it was not produced by any of the four transformations.
4	B	In the second choice, the larger object was reflected across the horizontal axis and dilated.
5	C	In order to have two similar objects, we must have a dilation.
6	A	One condition of similarity is that all dimensions are proportional.

Question No.	Answer	Detailed Explanation
7	C	The first transformation rotated the point 90° clockwise and the second translated it 3 units to the left.
8	A	AB : WX = 5 : 10 = 1 : 2 BC : XY = 3 : 6 = 1 : 2 AC : WY = 6 : 12 = 1 : 2
9	B	The ratio of the sides in rectangle B is 2 : 3, but the ratios of the sides in the other three rectangles is 1 : 2.
10	C	AB : WX = AC : WY 9 : 27 = 14 : WY 3 : 9 = 14 : WY 3(WY) = 9(14) WY = 42

Interior & Exterior Angles in Geometric Figures (8.G.A.5)

1	D	The statement in the problem is the definition of vertical angles.
2	C	a + b + c = 180 100 + c = 180 c = 80°
3	C	If two parallel lines are cut by a transversal, the corresponding angles are congruent.
4	B	If two parallel lines are cut by a transversal, the alternate interior angles are congruent.
5	C	If two parallel lines are cut by a transversal, the exterior angles on the same side of the transversal are supplementary.
6	D	If two parallel lines are cut by a transversal the alternate interior angles are congruent; so the measures of angle b and angle c are equal. If two straight lines intersect, the vertical angles are congruent; so the measures of angle a and angle c are equal. That makes the measures of angles a, b, and c all equal.
7	B	a + b + c = 180° 65° + 65° + c = 180° 130° + c = 180° c = 50°

Question No.	Answer	Detailed Explanation
8	B	$a + b + c = 180°$ $a + 40° + 30° = 180°$ $a = 110°$ $a + d = 180°$ $110° + d = 180°$ $d = 70°$ Also, the measure of an exterior angle of a triangle is equal to the sum of the measures of the two non-adjacent interior angles.
9	B	An interior angle of a triangle and its exterior angle are supplementary.
10	C	If two parallel lines are cut by a transversal, the measures of the alternate interior angles are equal.

Verifying the Pythagorean Theorem (8.G.B.6)

1	C	$3^2+4^2=5^2$ $9+16=25$ $25=25$
2	B	$8^2+15^2=c^2$ $64+225=c^2$ $289=c^2$ 17 cm $= c$
3	C	$7^2+11^2=c^2$ $49+121=c^2$ $170=c^2$ $13.0 \approx c$
4	C	$5^2+5^2=c^2$ $25+25=c^2$ $50=c^2$ 7 inches $\approx c$
5	D	$a^2+b^2=c^2$ cannot be changed to $a^2+c^2=b^2$.
6	C	$9^2+12^2=15^2$ $81+144=225$ $225=225$
7	A	$4^2+4^2= c^2$ $32 = c^2$ 5.7 in $\approx c$

Question No.	Answer	Detailed Explanation
8	B	Draw the figure. $16+a^2 = 100$ $a^2 = 84$ $a \approx 9.2$
9	C	$AC^2+BC^2=AB^2$ $AC^2+28^2=35^2$ $AC^2=1225-784$ $AC^2=441$ $AC=21$
10	D	$2s^2=625$ $s^2=312.5$ s is approximately 18

Pythagorean Theorem in Real-World Problems (8.G.B.7)

1	C	$8^2+h^2=17^2$ $h^2=289-64$ $h^2=225$ $h=15$ feet
2	A	Applying the Pythagorean Theorem to this (w, 18, 20) right triangle, $w^2 + 18^2 = 20^2$ is the correct equation.
3	D	$5^2+3^2=$(diagonal of bottom of chest)^2 $\sqrt{34}=$diagonal of bottom of chest $34+9=d^2$ $43=d^2$ $\sqrt{43}$ ft $=d$
4	A	Remember the Pythagorean triples (3,4,5 and their multiples). In this case: 30, 40, 50
5	B	$9^2+20^2=c^2$ $81+400=c^2$ $481=c^2$ $\sqrt{481}=c$ c is approximately 22 ft
6	B	$a^2+a^2=50^2$ $2a^2=2500$ $a^2=1250$ $a=35$ yd (approx)

Question No.	Answer	Detailed Explanation
7	A	$30^2+20^2=d^2$ $1300=d^2$ $\sqrt{1300}=d$ d=approx. 36 ft
8	B	$20^2+15^2=d^2$ $625=d^2$ 25 ft=d 20+15=35 ft The diagonal would be 10 ft shorter.
9	C	Use the Pythagorean triple: 30, 40, 50 He can place his ladder up to 30 feet away.
10	D	$25^2+x^2=50^2$ $x^2=2500-625=1875$ x=43 ft (approx)

Pythagorean Theorem & Coordinate System (8G.B.8)

1	A	$4^2+7^2=d^2$ $65=d^2$ d is approx. 8 m
2	C	Draw the graph. Draw the right triangle with the distance between the points as the hypotenuse. The legs will be 6 units and 8 units. Find the hypotenuse. $8^2+6^2=h^2$ $64+36=h^2$ $100=h^2$ 10=h
3	B	$4^2+14^2=d^2$ $16+196=d^2$ $212=d^2$ d=14.6
4	C	$25+9=34=d^2$ d=5.8
5	C	$4^2+4^2=d^2$ $32=d^2$ d=5.7

Question No.	Answer	Detailed Explanation
6	B	$2^2+10^2=AB^2$ $\sqrt{104}=AB$ $10.2=AB$ $4^2+9^2=AC^2$ $\sqrt{97}=AC$ $9.8=AC$
7	A	$50^2+40^2=d^2$ $2500+1600=d^2$ $4100=d^2$ 64 yd is approx. distance
8	B	Original point: (7,8) New point: (8,-7) Final point: (-8,-7) $15^2+15^2=d^2$ $450=d^2$ $d=21.2$ units
9	B	Draw a sketch. The coordinates of the rectangle are (0,0), (15,0), (15,20), and (0,20). Halfway across the width is 7.5 and halfway up the length is 10. Therefore, the center is at (7.5,10).
10	B	$3^2+3^2=d^2$ $18=d^2$ $4.2=d$

Finding Volume: Cone, Cylinder, & Sphere (8.G.C.9)

1	D	$V=4/3\,(\pi r^3)$ $V=4/3\,(\pi)\,6^3$ $V=4/3\,(216)\,\pi$ $V=288\pi$
2	C	$V=(1/3)Bh$ where B= area of the base $V=(1/3)\pi(4^2)(9)$ $V=48\pi$
3	C	$V=\pi r^2 h$ $V=\pi\,(25)(3)$ $V=75\pi$
4	D	$V=(4/3)\pi r^3$ $V=4/3(125\pi)$ $V=(500/3)\pi$ $V=523.6$ cubic inches

Question No.	Answer	Detailed Explanation
5	B	$V=\pi r^2 h$ $V=(4)(5)^2\pi$ $V=100\pi$ $V = 100(3.14)$ $V=314$ cu in
6	D	$V_{cylinder} = \pi r^2 h$ $V_{cone}=1/3\ \pi r^2 h$ If volumes are the same, then the height of the cone must be 3 times the height of the cylinder because $3(1/3)=1$.
7	C	$V_{sphere}=4/3\ \pi r^3$ $V=4/3\ (8)\pi$ $V=32/3\ \pi$ $V_{cylinder}=\pi r^2 h$ $V=8\pi$ $V_{cone}=1/3\ \pi r^2 h$ $V=(1/3)(9)(4)\pi$ $V=12\pi$; The cone has the greatest volume.
8	D	$V_{sphere}=4/3\ \pi r^3$ $V=4/3\ \pi(27)$ $V=36\pi$ $V_{cylinder}=\pi r^2 h$ $V=9(4)\pi$ $V=36\pi$ $V_{cone}=1/3\ \pi r^2 h$ $V=(1/3)(36)(3)\pi$ $V=36\pi$
9	A	$V_{sphere}/V_{cone}= 4/3\ \pi r^3 / 1/3\ \pi r^2 h = 4r/h = 4r/4 = r/1$
10	A	$V_{cylinder}/V_{cone}= \pi r^2 h/(1/3)\pi r^2 h=3/1$

Statistics and Probability

Interpreting Data Tables & Scatter Plots (8.SP.A.1)

1. If a scatter plot has a line of best fit that decreases from left to right, which of the following terms best describes the association?

   Ⓐ Positive association
   Ⓑ Negative association
   Ⓒ Constant association
   Ⓓ Nonlinear association

2. If a scatter plot has a line of best fit that increases from left to right, which of the following terms best describes the association?

   Ⓐ Positive association
   Ⓑ Negative association
   Ⓒ Constant association
   Ⓓ Nonlinear association

3. Which of the following scatter plots is the best example of a linear association?

Ⓐ

Ⓑ

Ⓒ

Ⓓ

4. Data for 9 kids' History and English grades are made available in the chart. What is the association between the History and English grades?

Kids	1	2	3	4	5	6	7	8	9
History	63	49	84	33	55	23	71	62	41
English	67	69	82	32	59	26	73	62	39

Ⓐ Positive association
Ⓑ Negative association
Ⓒ Nonlinear association
Ⓓ Constant association

5. Data for 9 kids' History grades and the distance they live from school are made available in the chart. What is the association between these two categories?

Kids	1	2	3	4	5	6	7	8	9
History	63	49	84	33	55	23	71	62	41
Distance from School (miles)	.5	7	3	4	5	2	3	6	9

Ⓐ No association
Ⓑ Positive association
Ⓒ Negative association
Ⓓ Constant association

6. Data for 9 kids' Math and Science grades are made available in the chart. What is the association between the Math and Science grades?

Kids	1	2	3	4	5	6	7	8	9
Science	63	49	84	33	55	23	71	62	41
Math	67	69	82	32	59	26	73	62	39

Ⓐ Positive association
Ⓑ No association
Ⓒ Constant association
Ⓓ Negative association

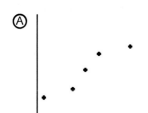
7. **Which of the scatter plots below is the best example of positive association?**

Ⓐ

Ⓑ

Ⓒ

Ⓓ

8. **Abbey hypothesized that the taller a person is the faster he or she will run. Is Abbey's hypothesis supported by the scatterplot below?**

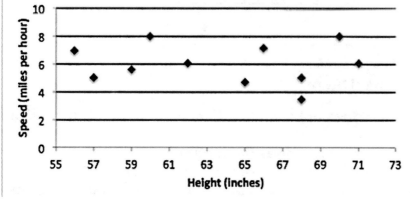

Ⓐ Yes, the taller people run faster.
Ⓑ Yes, taller people have longer legs and can cover more distance per stride.
Ⓒ No, the slowest person was relatively tall.
Ⓓ No, the line of best fit does not show a positive correlation between height and speed.

9. Jonathan conducted a science experiment to determine the relationship between the number of times his pet cricket would chirp and the temperature outside. Describe the correlation based on the scatterplot below.

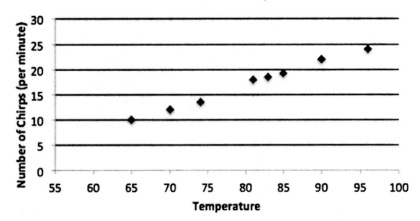

Ⓐ The hotter the temperature, the less energy the crickets have to chirp.
Ⓑ As the temperature increases, so does the number of chirps.
Ⓒ As temperature decreases, the number of chirps increases.
Ⓓ There is no relationship between temperature and number of chirps.

10. Everytime Damien buys a new video game, he invites his friends over to play. The scatterplot below graphs the number of friends he invites with the number of turns he has per hour. What is the relationship between these two variables?

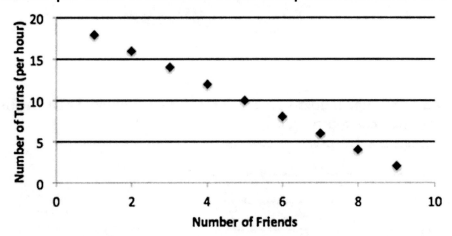

Ⓐ The number of friends invited does not seem to affect the number of turns Damien gets to play.
Ⓑ As the number of friends increases, so does the number of turns.
Ⓒ As the number of friends increases, the number of turns decreases.
Ⓓ There is no relationship between the number of friends invited and the number of turns.

Scatter Plots, Line of Best Fit (8.SP.A.2)

1.

Which of the following best describes the points in this scatter plot?

Ⓐ Increasing Linear
Ⓑ Decreasing Linear
Ⓒ Constant Linear
Ⓓ None of these

2.

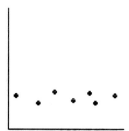

Which of the following best describes the points in this scatter plot?

Ⓐ Increasing Linear
Ⓑ Decreasing Linear
Ⓒ Constant Linear
Ⓓ None of these

3.

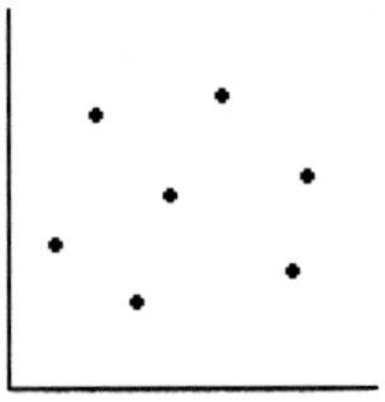

Which of the following lines best approximates the data in the scatter plot shown above?

Ⓐ

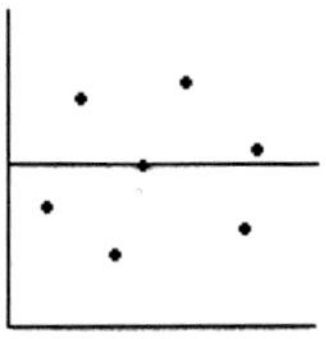

Ⓑ

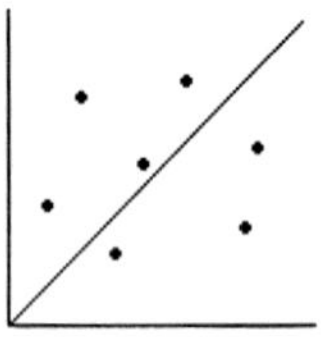

Ⓒ

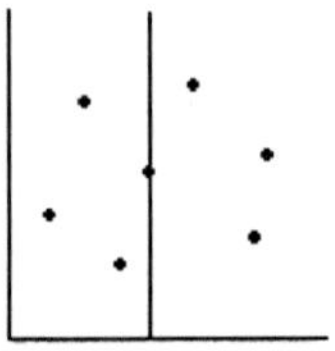

Ⓓ None of these; the data do not appear to be related linearly.

4. Which scatter plot represents a positive linear association?

Ⓐ

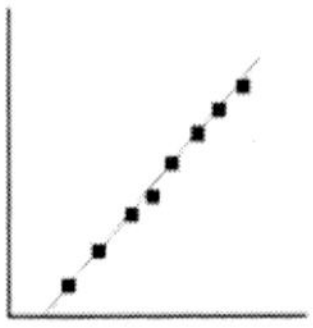

Ⓑ

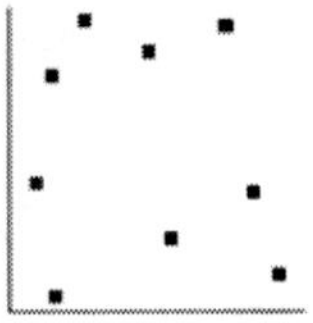

Ⓒ

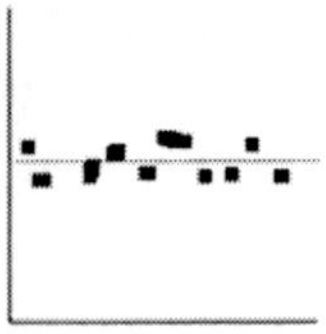

Ⓓ

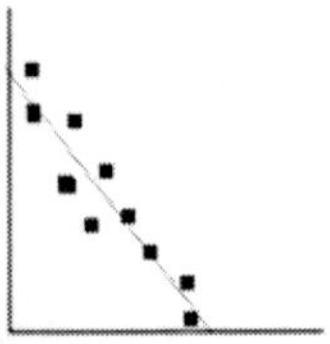

5. **Which scatter plot represents a negative linear association?**

Ⓐ

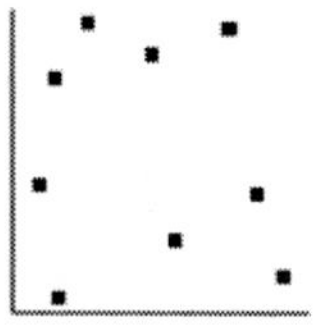

Ⓑ

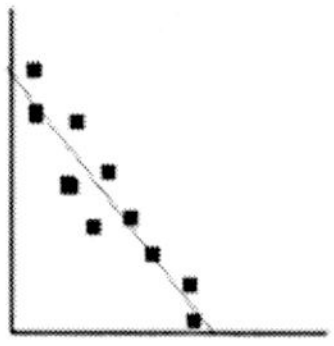

Ⓒ

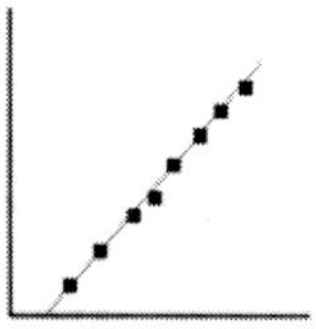

Ⓓ

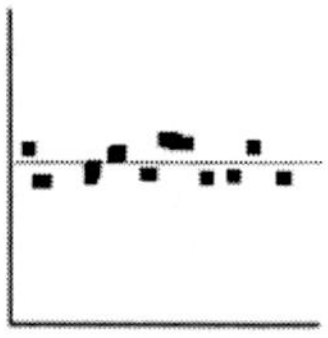

6. **Which scatter plot represents no association?**

Ⓐ

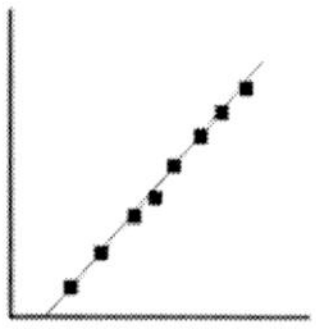

Ⓑ

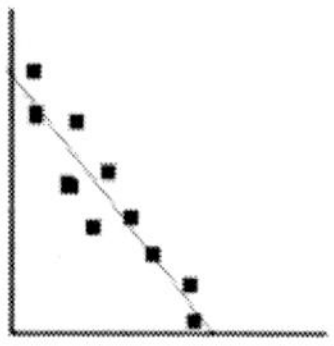

Ⓒ

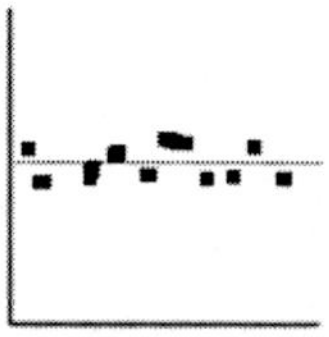

Ⓓ

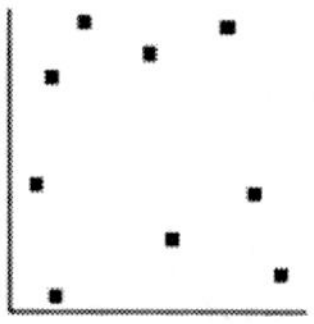

7. **Which scatter plot represents a constant association?**

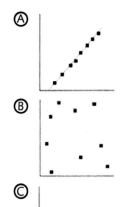

8. **The graph of this data set would most resemble which of the following graphs?**

x	1	2	3	4	5	6	7
y	2	3	4	5	6	7	8

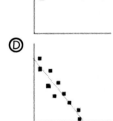

9. The graph of this data would most resemble which of the following graphs?

x	1	2	3	4	5	6	7
y	4	4	4	4	4	4	4

Ⓐ

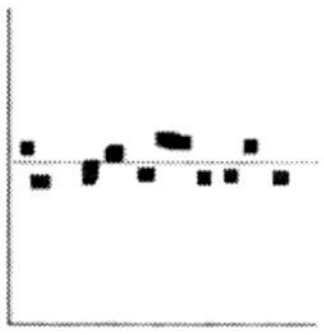

Ⓑ

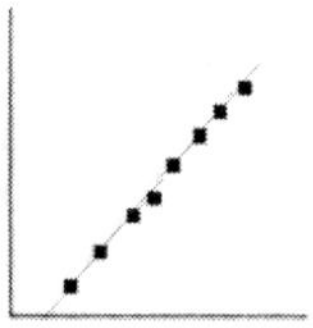

Ⓒ

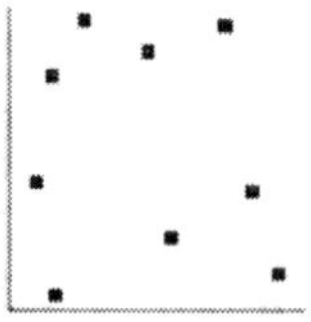

Ⓓ

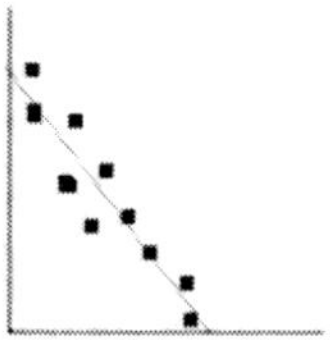

10. Typically air temperature decreases through the night between midnight and 6:00 am. This is an example of what type of association?

Ⓐ constant association
Ⓑ positive linear association
Ⓒ negative linear association
Ⓓ no association

Analyzing Linear Scatter Plots (8.SP.A.3)

1. Students in Mrs. Dee's class planted seeds in the school garden. Every student was allowed to choose how many hours of sun exposure their plant received each day. The scatter plot below plots the number of hours of sunlight the plant received and its final height.

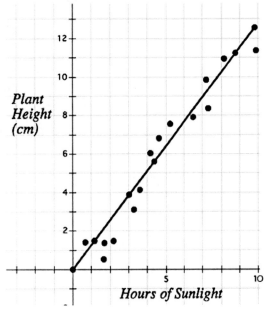

How much did the height of the plant increase for each hour of sun exposure?

Ⓐ $\frac{3}{4}$ cm

Ⓑ 1 cm

Ⓒ $1\frac{1}{3}$ cm

Ⓓ 2 cm

2. Marcella noticed that the more time she spent getting ready in the morning the more compliments she received throughout the day. The scatter plot compares the number of minutes to the number of compliments.

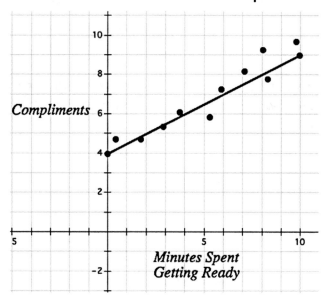

How many compliments will Marcella receive if she spends no time getting ready in the morning?

Ⓐ 0 compliments
Ⓑ 2 compliments
Ⓒ 4 compliments
Ⓓ 10 compliments

3. Marcella noticed that the more time she spent getting ready in the morning the more compliments she received throughout the day. The scatter plot compares the number of minutes to the number of compliments.

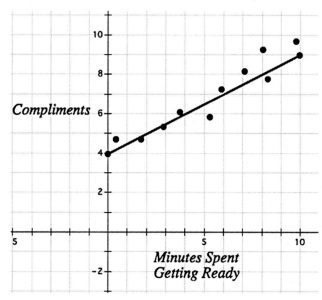

How much do compliments increase for each minute spent getting ready?

Ⓐ .5 compliments/minute
Ⓑ 1 compliment/minute
Ⓒ 2 compliments/minute
Ⓓ 4 compliments/minute

4. Marcella noticed that the more time she spent getting ready in the morning the more compliments she received throughout the day. The scatterplot compares the number of minutes to the number of compliments.

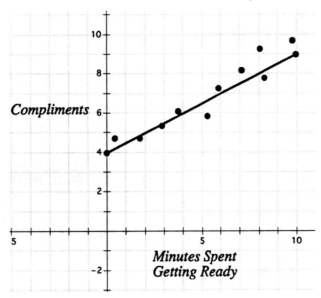

Marcella would like to receive fifteen compliments today. Use the line of best fit to determine how many minutes she must spend getting ready.

Ⓐ 18 minutes
Ⓑ 22 minutes
Ⓒ 26 minutes
Ⓓ 30 minutes

5. Jamie bought a used car for $12,500. He read that the value of the car was likely to decrease by $1,500 each year. The scatterplot below shows the value of Jamie's car over time.

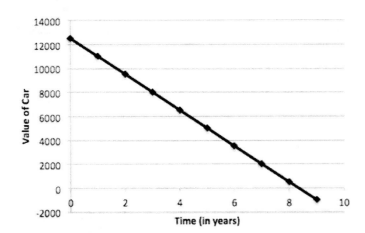

What is the equation of the best fit line?

Ⓐ y = -12,500x -1,500
Ⓑ y = -12,500x +1,500
Ⓒ y = -1,500x +12,500
Ⓓ y = -1,500x -12,500

6. Jamie bought a used car for $12,500. He read that the value of the car was likely to decrease by $1,500 each year. The scatterplot below shows the value of Jamie's car over time.

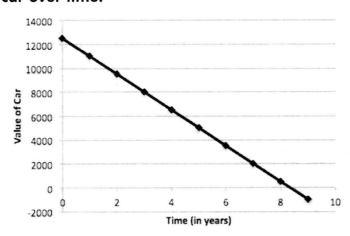

After about how much time will Jamie's car be worth nothing?

Ⓐ 7.5 years
Ⓑ 8.3 years
Ⓒ 8.8 years
Ⓓ 9.6 years

7. Jamie bought a used car for $12,500. He read that the value of the car was likely to decrease by $1,500 each year. The scatterplot below shows the value of Jamie's car over time.

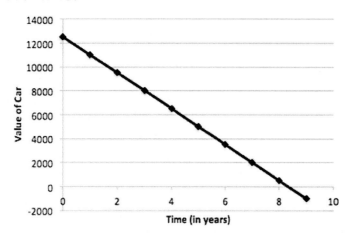

Why is the y-intercept at 12,500?

Ⓐ This is the original value of the car.
Ⓑ This is the amount he wanted to pay.
Ⓒ This is the amount he pays each month.
Ⓓ This is the amount he could see his car for after 1 year.

8. Student government wants to sell cookies as a fundraiser. Based on past cookie sales, the student government anticipates that if cookies are priced at $3 then 96 will sell, and if cookies are priced at $2 then 300 will sell. The graph below plots both these values.

How much do cookie sales decrease for each $1 increase in price?

Ⓐ 32 cookies
Ⓑ 66 cookies
Ⓒ 150 cookies
Ⓓ 204 cookies

9. Student government wants to sell cookies as a fundraiser. Based on past cookie sales, the student government anticipates that if cookies are priced at $3 then 96 will sell, and if cookies are priced at $2 then 300 will sell. The graph below plots both these values.

What is the equation for the line of best fit?

Ⓐ y = -204 x + 708
Ⓑ y = -104x + 708
Ⓒ y = -204x + 504
Ⓓ y = -96x + 504

10. Student government wants to sell cookies as a fundraiser. Based on past cookie sales, the student government anticipates that if cookies are priced at $3 then 96 will sell, and if cookies are priced at $2 then 300 will sell. The graph below plots both these values.

Based on this graph, if the cookies are priced at $2.50, how many cookies should student government expect to sell?

Ⓐ 168 cookies
Ⓑ 172 cookies
Ⓒ 198 cookies
Ⓓ 204 cookies

1. In a collection of 80 books, 34 have leather covers and 18 were published before 1850. If 42 of these books were published after 1850 and do not have leather covers, how many books were published before 1850 and have leather covers?

    Ⓐ 9
    Ⓑ 4
    Ⓒ 14
    Ⓓ 20

2.

	Plays an instrument	Does not play an instrument
Plays a Sport	60	30
Does not play a sport	10	50

150 students were surveyed and asked whether they played a sport and whether they played a musical instrument. The results are shown in the table above.
Which two sections add up to just over half of the number of students surveyed?

Ⓐ The two sections that do not play an instrument.
Ⓑ The two sections that do not play a sport.
Ⓒ The two sections that play an instrument.
Ⓓ The two sections that play a sport.

3. The chart below shows how many of the students in a particular school participated in extra-curricular activities and whether they were listed on the honor roll.

Honor Roll vs Extra Curricular

	Extra Curricular	Not EC
Honor Roll	83	79
Not HR	167	54

Of the total number of students researched, what percent made the honor roll?

Ⓐ 42%
Ⓑ 50%
Ⓒ 67%
Ⓓ None of the above.

4.

	Fertilizer	No Fertilizer
Lived	200	600
Died	50	150

Out of 1,000 plants, some were given a new fertilizer and the rest were given no fertilizer. Some of the plants lived and some of them died, as shown in the table above. Which of the following statements is supported by the data?

Ⓐ Fertilized plants died at a higher rate than unfertilized plants did.
Ⓑ Fertilized plants and unfertilized plants died at the same rate.
Ⓒ Fertilized plants died at a lower rate than unfertilized plants died.
Ⓓ None of the above statements can be supported by the data.

5.

	Windy	Not Windy
Sunny	5	15
Cloudy	4	6

The weather was observed for 30 days; each day was classified as sunny or cloudy, and also classified as windy or not windy. The results are shown in the table above. Which of the following statements is NOT supported by the data?

Ⓐ 25% of the sunny days were also windy.
Ⓑ 30% of the days were windy.
Ⓒ 40% of the cloudy days were also windy.
Ⓓ 50% of the windy days were also sunny.

6.

	Jeans	No Jeans
Sneakers	15	10
No Sneakers	5	20

50 people were asked whether they were wearing jeans and whether they were wearing sneakers. The results are shown in the table above.
What fraction of the people who wore sneakers were NOT wearing jeans?

Ⓐ $\frac{1}{5}$

Ⓑ $\frac{2}{5}$

Ⓒ $\frac{3}{10}$

Ⓓ $\frac{3}{4}$

7.

	Jeans	No Jeans
Sneakers	15	10
No Sneakers	5	20

50 people were asked whether they were wearing jeans and whether hey were wearing sneakers. The results are shown in the table above. Which of the following statements is NOT supported by the data?

Ⓐ A randomly chosen person who is wearing sneakers is equally as likely to be wearing jeans as not wearing jeans.

Ⓑ A randomly chosen person who is not wearing jeans is 2 times as likely to be not wearing sneakers as wearing sneakers.

Ⓒ A randomly chosen person who is wearing jeans is 3 times as likely to be wearing sneakers as not wearing sneakers.

Ⓓ A randomly chosen person who is not wearing sneakers is 4 times as likely to be not wearing jeans as wearing jeans.

8.

	Enjoys Crosswords	Does not Enjoy Crosswords
Enjoys Sudoku	30	20
Does not Enjoy Sudoku	40	10

100 people were asked whether they enjoy crossword puzzles and whether they enjoy sudoku number puzzles. The results are shown in the table above.

What percent of all 100 people enjoy sudoku?

Ⓐ 20%
Ⓑ 30%
Ⓒ 50%
Ⓓ 60%

9.

	Enjoys Crosswords	Does not Enjoy Crosswords
Enjoys Sudoku	30	20
Does not Enjoy Sudoku	40	10

100 people were asked whether they enjoy crossword puzzles and whether they enjoy sudoku number puzzles. The results are shown in the table above.
Which of the following statements is NOT supported by the data?

Ⓐ A randomly chosen person who enjoys sudoku is more likely to enjoy crosswords than to not enjoy crosswords.
Ⓑ A randomly chosen person who does not enjoy sudoku is more likely to enjoy crosswords than to not enjoy crosswords.
Ⓒ A randomly chosen person who enjoys crosswords is more likely to enjoy sudoku than to not enjoy sudoku.
Ⓓ A randomly chosen person who does not enjoy crosswords is more likely to enjoy sudoku than to not enjoy sudoku.

10.

Preferred Sports

	Volleyball	Basketball	Softball
Boys	5	30	15
Girls	30	5	15

Out of those students who preferred volleyball, about what percent were girls?

Ⓐ 15%
Ⓑ 35%
Ⓒ 85%
Ⓓ 100%

End of Statistics and Probability

Answer Key and Detailed Explanations
Statistics and Probability
Interpreting Data Tables & Scatter Plots (8.SP.A.1)

Question No.	Answer	Detailed Explanation
1	B	By definition, a decreasing trend from left to right on a scatter plot indicates a negative association.
2	A	By definition, an increasing line from left to right on a scatter plot indicates a positive association.
3	C	The points in the third choice are nearly in a straight line.
4	A	As the History grade increases, so does the English grade. Thus, there is a positive association.
5	A	There does not appear to be any significant correlation between these two variables.
6	A	As the Science grade increases, so does the Math grade. This indicates a positive association.
7	A	In the first graph, a best fit line would be nearly a straight line increasing to the right.
8	D	The line of best fit would be a horizontal line, which would not indicate a positive correlation.
9	B	The line of best fit for this data has a positive slope, indicating a positive correlation between temperature and number of cricket chirps.
10	C	The line of best fit for this data has a negative slope, indicating an inverse correlation between number of friends and number of turns.

Scatter Plots, Line of Best Fit (8.SP.A.2)

1	D	This scatter plot does not represent a linear function. None of these is the correct choice.
2	C	The line of best fit would be a horizontal line. Constant linear is the correct choice.
3	D	The data does not represent a straight line. None of these is the correct choice.
4	A	The first scatter plot shows a positive slope representing a positive linear association.
5	B	The second scatter plot shows a negative slope representing a negative linear association.

Question No.	Answer	Detailed Explanation
6	D	The data on the fourth scatter plot cannot be represented linearly and, therefore, represents no association.
7	C	The third scatter plot represents a horizontal line or a constant association.
8	B	The data represents a straight line with a slope of positive 1. The second scatter plot is the correct choice.
9	A	The y-coordinate is a constant 4 representing a horizontal line at y=4. The first scatter plot is the correct choice.
10	C	Temperature decreases as the hours increase. Negative linear association is the correct choice.

Analyzing Linear Scatter Plots (8.SP.A.3)

1	C	The slope of the line is $\frac{4}{3}$ or $1\frac{1}{3}$, indicating $1\frac{1}{3}$ centimeters of growth per 1 hour of sunlight.
2	C	The y-intercept is 4. Therefore with 0 minutes spent getting ready, Marcella will still receive 4 compliments.
3	A	The slope of the line is $\frac{1}{2}$ or .5, indicating that she will receive .5 compliment for every 1 minute spent getting ready.
4	B	Given the y-intercept and slope from the graph, the equation for the line of best fit is y=.5x+4. By substituting y=15, (15)=.5x+4, x=22
5	C	Since the original value represents the y-intercept, the y-intercept is 12,500. The rate of change is decreasing $1500 per year, thus m=-1500. Therefore, using y=mx+b, the equation representing this situation
6	B	Given that the equation of the line is Y = -1,500x + 12,500, substitute Y = 0. 0 = -1,500x + 12,500 - 12,500 = -1,500x 8.3 ≈ x
7	A	The y-intercept represents the starting value (after zero time has elapsed) or in this case the original amount of the car.

Question No.	Answer	Detailed Explanation
8	D	The amount of decrease is represented by the slope of the line. Since two points are given, the slope is $$\frac{\triangle y}{\triangle x} = \frac{96 - 300}{3 - 2} = -204$$
9	A	Given that the slope of the line is m=-204, substitute one of known points to solve for b. $(300) = -204(2) + b$ $300 = -408 + b$ $708 = b$ $y = -204x + 708$
10	C	Given the equation of the line is $y = -204x + 708$, substitute x=2.5. $y = -204(2.5) + 708$ $y = -510 + 708$ $y = 198$

Relatable Data Frequency (8.SP.A.4)

Question No.	Answer	Detailed Explanation
1	C	Book Binding vs Publication Dates

	Books	Before 1850	After 1850
Leather	34	14	20
Not Leather	46	4	42

Question No.	Answer	Detailed Explanation
2	A	See table. 30+50=80, which is about half of 150.
3	A	$\frac{162}{383} = 42.3\%$
4	B	Fertilized plants: 50/200 =1/4 died Unfertilized plants: 150/600=1/4 died They died at the same rate is the correct choice.
5	D	There were a total of 9 windy days and 5 of them were also sunny. 5/9=56% The correct choice is 50% of the windy days were also sunny.

Question No.	Answer	Detailed Explanation
6	B	Out of 25 people wearing sneakers, 10 were not wearing jeans. 10/25 = 2/5 $\dfrac{2}{5}$ is the correct answer.
7	A	Out of 25 people wearing sneakers, 15 were wearing jeans and 10 were not. The statement,"A randomly chosen person who is wearing sneakers is equally as likely to be wearing jeans as not wearing jeans," is not supported by the data.
8	C	50 out of 100 people enjoy sudoku. The correct answer is 50%.
9	C	Out of 70 people who enjoy crosswords, 30 enjoy sudoku and 40 do not. The statement, "A randomly chosen person who enjoys crosswords is more likely to enjoy sudoku than to not enjoy sudoku," is not supported by the data.
10	C	Out of 35 students who preferred volleyball, 30 were girls. 30/35 = 86% 85% is the correct choice.

Notes

Lumos StepUp™ is an educational app that helps students learn and master grade-level skills in Math and English Language Arts.

The list of features includes:

- Learn Anywhere, Anytime!

- Grades 3-8 Mathematics and English Language Arts

- Get instant access to the Common Core State Standards

- One full-length sample practice test in all Grades and Subjects

- Full-length Practice Tests, Partial Tests and Standards-based Tests

- 2 Test Modes: Normal mode and Learning mode

- Learning Mode gives the user a step-by-step explanation if the answer is wrong

- Access to Online Workbooks

- Provides ability to directly scan QR Codes

- And it's completely FREE!

http://lumoslearning.com/a/stepup-app

About Online Workbooks

- ◆ When you buy this book, 1 year access to online workbooks is included

- ◆ Access them anytime from a computer with an internet connection

- ◆ Adheres to the Common Core State Standards

- ◆ Includes progress reports

- ◆ Instant feedback and self-paced

- ◆ Ability to review incorrect answers

- ◆ Parents and Teachers can assist in student's learning by reviewing their areas of difficulty

Course Name: Grade 4 Math Prep

Lesson Name:	Correct	Total	% Score	Incorrect
Introduction				
Diagnostic Test		3	0%	3
Number and Numerical Operations				
Workbook - Number Sense	2	10	20%	8
Workbook - Numerical Operations	2	25	8%	23
Workbook - Estimation	1	3	33%	2
Geometry and measurement				
Workbook - Geometric Properties		6	0%	6
Workbook - Transforming Shapes				
Workbook - Coordinate Geometry	1	3	33%	2
Workbook - Units of Measurement				
Workbook - Measuring Geometric Objects	3	10	30%	7
Patterns and algebra				
Workbook - Patterns	7	10	70%	3
Workbook - Functions and relationships				

LESSON NAME: Workbook - Geometric Properties

Elapsed Time: 01:19

Question No. 2

What type of motion is being modeled here?

Select right answer
- ◯ a translation
- ◯ a rotation 90° clockwise
- ◉ a rotation 90° counter-clockwise
- ◯ a reflection

[Previous question]   [Next question]

Report Name: Missed Questions

Student Name: Lisa Colbright
Cours Name: Grade 4 Math Prep
Lesson Name: Diagnostic Test

The faces on a number cube are labeled with the numbers 1 through 6. What is the probability of rolling a number greater than 4?

Answer Explanation

(C) On a standard number cube, there are six possible outcomes. Of those outcomes, 2 of them are greater than 4. Thus, the probability of rolling a number greater than 4 is "2 out of 6" or 2/6.

A)		1/6
B)		1/3
C)	Correct Answer	2/6
D)		3/6

Lumos Learning
Developed By Expert Teachers

Grade **8**

Common Core
Practice
ENGLISH

(((tedBook)))

LANGUAGE ARTS

⭐ Three Strands

⭐ Hundreds of Activities

30+

SKILLS

PLUS **Online Workbooks**

Foundational Skills for
PARCC or
Smarter Balanced Tests

Available

- At Leading book stores
- Online www.LumosLearning.com

CPSIA information can be obtained
at www.ICGtesting.com
Printed in the USA
LVOW09s1644061117
555217LV00008B/730/P